Reworking Qualitative Data

Reworking Qualitative Data

Janet Heaton

SAGE Publications
London • Thousand Oaks • New Delhi

© Janet Heaton 2004

First published 2004

All rights reserved. No part of this publication may be reproduced, stored in a retrieval system, transmitted or utilized in any form or by any means, electronic, mechanical, photocopying, recording or otherwise, without permission in writing from the Publishers.

SAGE Publications Ltd
1 Oliver's Yard
55 City Road
London EC1Y 1SP

SAGE Publications Inc
2455 Teller Road
Thousand Oaks, California 91320

SAGE Publications India Pvt Ltd
B-42, Panchsheel Enclave
Post Box 4109
New Delhi 100 017

British Library Cataloguing in Publication data

A catalogue record for this book is available
from the British Library

ISBN 0-7619-7142-4
    0-7619-7143-2 (pbk)

Library of Congress control number available

Typeset by C&M Digitals (P) Ltd, Chennai, India
Printed and bound in Great Britain by Athenaeum Press, Gateshead

Contents

List of tables   vii

Preface   viii

Acknowledgements   xi

1 What is Secondary Analysis?   1
    The analysis of pre-existing data in social research   2
    Functions of secondary analysis   8
    Modes of secondary analysis   12
    Conclusion   15
    Notes   16

2 From Quantitative to Qualitative Secondary Analysis   19
    Developments in data archiving and computing   20
    The promotion of data sharing and data retention   22
    Potential benefits and problems of re-using data   26
    Conclusion   31
    Notes   32

3 Types of Qualitative Secondary Analysis   35
    Overview of the studies   36
    Supra analysis   39
    Supplementary analysis   41
    Re-analysis   45
    Amplified analysis   47
    Assorted analysis   50
    Conclusion   51
    Notes   52

4 Epistemological Issues   54
    The nature of qualitative research   54
    The problem of data 'fit'   57
    The problem of not having 'been there'   60
    The problem of verification   65
    Conclusion   71
    Notes   71

5 Ethical and Legal Issues   73
    Ethical and legal frameworks for social research   74
    Informed consent   77
    Confidentiality   81
    Copyright   83

Contents

	Data protection	85
	Defamation	85
	Conclusion	86
	Notes	86
6	*Modi Operandi*	89
	Design	90
	Selection of data set(s)	91
	Analysis	96
	Quality assurance	99
	Reportage	102
	Reflexivity	104
	Conclusion	106
	Notes	107
7	**The Future of Qualitative Secondary Analysis**	109
	The nature of qualitative secondary analysis	110
	Secondary analysis and the 'landscape' of qualitative research	112
	The secondary analyst as *bricoleur*	116
	Resources for qualitative secondary analysis	119
	Conclusion	124
	Notes	124

Appendix A	The Literature Review	126
Appendix B	Criteria for the Evaluation of Qualitative Research Papers	131

Bibliography	135
Key Names and Titles Index	151
Subject Index	156

List of tables

Table 1.1	Examples of pre-existing quantitative data used in social research	3
Table 1.2	Examples of pre-existing qualitative data used in social research	5
Table 1.3	Modes of secondary analysis	12
Table 2.1	Recommendations of the Committee on National Statistics	24
Table 3.1	Authors' relationship to the primary data set(s)	37
Table 3.2	Types of secondary analysis of qualitative data	38
Table 4.1	Axioms of positivist and naturalistic inquiry	56
Table 6.1	Guidelines for assessing the re-usability of qualitative data sets	93
Table 6.2	Examples of possible methodological matters to be described in reports of qualitative secondary studies	104
Table 7.1	The seven moments of qualitative research	113
Table A1.1	Key terms used in the literature search	127
Table A1.2	Aspects of studies appraised in the review	128
Table A1.3	Principal characteristics of types of secondary analysis	129

Preface

> To be sure, secondary analysis is not limited to quantitative data. Observation notes, unstructured interviews and documents can also be usefully reanalyzed. In fact, some field workers may be delighted to have their notes, long buried in their files, reanalyzed from another point of view... Man is a data gathering animal. (Glaser, 1962: 74)

Barney Glaser was one of the first analysts to recognize the potential of re-using data from qualitative research studies. Better known for his work with Anselm Strauss on grounded theory, he published two articles in the 1960s on the secondary analysis of survey data in which he briefly discusses the possibilities of extending secondary analysis from quantitative to qualitative data (Glaser, 1962, 1963). Four decades later, however, secondary analysis remains ill-defined and underdeveloped as a qualitative methodology. Although Glaser's view has subsequently been echoed by others who claim that there '...is no reason why statistical data should be the only material available for secondary analysis' (Dale et al., 1988: 15), the secondary analysis of qualitative data remains an enigma.

But the picture is changing. Since the mid-1990s, researchers in North America and the United Kingdom have begun to explore in more depth the potential of secondary analysis as a qualitative methodology (see Heaton, 1998; Hinds et al., 1997; Szabo and Strang, 1997; Thorne, 1994, 1998). At the same time, the Qualitative Data Archival Resource Centre (Qualidata) was established in the United Kingdom in 1994 to promote the archiving of qualitative data from the social sciences (Corti and Thompson, 1998). Funded by the Economic and Social Research Council (ESRC), Qualidata was the world's first co-ordinated effort to facilitate the archiving of qualitative data sets on a national basis. The work of Qualidata has led to an increase in the number of qualitative data sets archived in the United Kingdom. It has also contributed to and helped to fuel a growing national and international debate over the possibilities and problems of re-using qualitative data and, in so doing, sought to stimulate a culture of data archiving and re-use in qualitative research in the social sciences (see Backhouse, 2002; Corti, 1999, 2000; Corti et al., 1995; Thompson, 2000a).

In this book, which is my own contribution to this debate as a qualitative researcher who is interested in both the sociology of knowledge and the utility of the methodology as a 'tool of the trade', I explore the making(s) of secondary analysis as a qualitative methodology. The book focuses on the question: what is the nature of qualitative secondary analysis and is it compatible with the key tenets of qualitative inquiry? It aims to explore the fundamentals of the methodology as a way of informing future practice involving the collection, retention and re-use of qualitative data. While it provides information and advice for researchers interested in doing qualitative secondary analysis, it is not intended as a manual as such, but more a resource on the principles of, and issues raised by, the re-use of qualitative data in social research. In these respects it echoes Herbert H. Hyman's (1972) early work on the secondary analysis of survey data.

The book is based on a detailed review of examples of qualitative secondary studies drawn mainly from the international health and social care literature, together with the small but growing theoretical and methodological literature on the re-use of qualitative data. The review was originally carried out with funding from the Economic and Social Research Council (Heaton, 2000) following earlier related work (Heaton, 1998), and has been updated for the present publication. Some secondary studies from other areas of social research have also been included to illustrate particular points. Although the focus is mainly on secondary studies drawn from health and social care research, this is an important area for three main reasons: first, it is the area in which key exponents of the methodology have pioneered the approach and explored its potential; secondly, it is one of the areas of social research where there is particular interest in the potential of qualitative secondary analysis and where a growing number of secondary studies have been published (other areas include education and criminology); and, finally, it is an area where ethical issues relating to the collection and re-use of sensitive qualitative data are paramount. Further details of the aims and conduct of the review are presented in Appendix A.

Overview of contents

I begin by examining existing definitions of quantitative and qualitative secondary analysis in Chapter 1 and exploring similarities and differences in the ways in which they have been conceptualized. At the same time, I consider how qualitative secondary analysis differs from other methodologies that also involve the re-use of qualitative data and evidence, such as documentary analysis, conversation analysis and meta-analysis or synthesis. These two themes – the distinction between qualitative and quantitative secondary analysis, and between qualitative secondary analysis and other qualitative methodologies – are developed in subsequent chapters.

Chapter 2 examines the history of secondary analysis and the extension of the methodology from quantitative to qualitative research data. Here I consider the factors that have both helped and hindered the development of qualitative secondary analysis. In particular, I examine the influence of advances in data archiving and computing, the rise of a data sharing imperative in the social sciences, and the debate over the pros and cons of re-using quantitative and qualitative data respectively.

In the next chapters, I examine existing examples of studies involving the re-use of qualitative data and explore how the methodology has been defined and developed in practice. Chapter 3 describes the characteristics of the studies that were reviewed and distinguishes five types of qualitative secondary analysis. The issues arising from these studies and the wider literature on the re-use of qualitative data are then discussed in the following three chapters. Chapter 4 focuses on the epistemological issues that were identified and, in particular, the problems of data 'fit', not 'being there' and verification. In Chapter 5, I examine the ethical and legal issues that were raised, including the matters of informed consent, confidentiality, copyright, data protection and defamation. Then, in Chapter 6, I explore what the process of re-using qualitative data involves in practice and consider the methodological implications for both primary and secondary qualitative researchers, as well as archivists. In the course of these three chapters, I introduce and elaborate the third major theme of the book, namely the diverse nature of qualitative secondary analysis itself.

In the last chapter, I reflect on the main findings of the review and return to the question of whether qualitative secondary analysis does have the makings of a qualitative methodology. Drawing on Denzin and Lincoln's (1994, 1998, 2000) analysis of the contemporary 'landscape' of qualitative research, I consider the extent to which the existing work of researchers who have re-used qualitative data resembles the approach of the metaphorical figure of the *bricoleur* expounded in their analysis, and the implications of this, and other styles of qualitative research, for the future development of qualitative secondary analysis. In conclusion, I highlight the resources that are needed to further develop and extend secondary analysis as a qualitative methodology.

Acknowledgements

The original literature review upon which the book is based was funded by the Economic and Social Research Council (Research Grants Scheme, reference: R0002222918). The book was written with the help of funding from the Innovation and Research Priming Fund (reference: 404) and the Social Policy Research Unit (SPRU) at the University of York.

This is my first book and I would like to thank all those who helped me to get started and see it through to the end. In particular, I would like to thank Tricia Sloper and Anne Corden from SPRU, and Andrew Tudor from the Department of Sociology at the University of York, for their help at various stages of the book's preparation. The book would have been a lesser work but for their input, and the support of others whom I have not singled out. Nonetheless, the views expressed are my own and I accept responsibility for any errors in and failings of the text.

1
What is Secondary Analysis?

The analysis of pre-existing data in social research	2
Functions of secondary analysis	8
Modes of secondary analysis	12
Conclusion	15
Notes	16

Secondary analysis is best known as a methodology for doing research using pre-existing statistical data. Social scientists in North America and Europe have been making use of this type of data throughout the twentieth century, although it was not until 1972 that the first major text on the research strategy, *Secondary Analysis of Sample Surveys: Principles, Procedures and Potentialities* by Herbert H. Hyman, was published. Since then, the literature on secondary analysis of quantitative data has grown considerably as the availability and use of these data has expanded. There is now a substantial body of work exploring different aspects of the methodology, including several textbooks describing the availability of statistical data sets and how they can be used for secondary research purposes (see Dale et al.,1988; Hakim, 1982; Kiecolt and Nathan, 1985; Stewart and Kamins, 1993), as well as critical commentaries on the scientific, ethical and legal aspects of sharing this type of data in the social sciences (see Fienberg et al., 1985; Hedrick, 1988; Sieber, 1988, 1991; Stanley and Stanley, 1988).[1] Accordingly, the terms 'secondary analysis', 'secondary data' and 'data sharing' have become synonymous with the re-use of statistical data sets.

However, in recent years interest has grown in the possibility of re-using data from qualitative studies. Since the mid-1990s a number of publications have appeared on the topic written by researchers who have carried out ground-breaking secondary analyses of qualitative data (Heaton, 1998;

Hinds et al., 1997; Mauthner et al., 1998; Szabo and Strang, 1997; Thompson, 2000a; Thorne, 1994, 1998), by archivists involved in the preservation of qualitative data sets for possible secondary analysis (see Corti et al., 1995; Corti and Thompson, 1998; Fink, 2000; James and Sørensen, 2000) and by academics interested in these developments (see Alderson, 1998; Hammersley, 1997a; Hood-Williams and Harrison, 1998). The extension of secondary analysis to qualitative data raises a number of questions about the nature of this research strategy. What is secondary analysis? How does the secondary analysis of qualitative data compare to that of quantitative data? And how is secondary analysis distinct from other quantitative and qualitative methodologies used in social research?

In this chapter, I explore these issues by comparing and contrasting the ways in which secondary analysis has been conceptualized as a methodology for re-using quantitative and qualitative data. I begin by examining the types of pre-existing data which are deemed to be subject to secondary analysis and related methodologies. This is followed by an outline of the purported functions of secondary analysis and how these are more – or less – accepted in relation to the re-use of quantitative and qualitative data. In the last section, I describe three modes of secondary analysis and the emphasis given to each in existing conceptualizations of quantitative and qualitative secondary analysis respectively. The chapter concludes with a provisional definition of secondary analysis which clarifies the focus of this book, as well as providing a basis for the subsequent exploration of the theoretical and substantive issues arising from this conceptual overhaul.

The analysis of pre-existing data in social research

Although the secondary analysis of quantitative data is an established methodology in social research, there is no single, unequivocal definition of the approach. Existing conceptualizations of quantitative and qualitative secondary analysis tend to refer to various propositions, some of which are more accepted than others. The first and most rudimentary principle of secondary analysis is that it involves the use of *pre-existing* data. This view is exemplified by the following who, writing on the secondary analysis of statistical data, claim that: 'Secondary analysis must, by definition, be an empirical exercise carried out on data that has already been gathered or compiled in some way' (Dale et al., 1988: 3). There are, however, notable differences in the types of pre-existing data which are subject to secondary analysis, depending on the nature and origin of the material.

Table 1.1 Examples of pre-existing quantitative data used in social research

Type	Examples
Census data	United States since 1790. England and Wales since 1801
Institutions' administrative data	Hospital records; employee records
Public records	Public Record Office; libraries
Social surveys (including omnibus and uni-purpose surveys by research centres and governments)	General Social Survey (GSS) in United States. British Social Attitudes Survey (BSAS) and General Household Survey (GHS) in Britain
Longitudinal studies	Terman database 1922 – present in United States. Longitudinal Study (LS); National Child Development Study (NCDS) and National Survey of Health and Development in Britain

Quantitative data

Pre-existing data used in quantitative secondary analysis has been derived from various activities, including research projects carried out by academics, government agencies and commercial groups, as well as the administrative work of public authorities and other organizations that routinely keep records for management purposes.[2] Table 1.1 shows examples of the kinds of statistical data that have been used in secondary research in the social sciences, derived from previous research studies and from other contexts.[3]

For early social researchers, the main types of pre-existing quantitative data available to them were census and administrative records. Durkheim (1952), for example, used both types of data in his classic study on suicide and Miller (1967) used administrative data in her study of former 'mental patients' adjustment to the world after discharge from hospital. As Hyman (1972) observes these resources were more difficult to use in the past. Nowadays census data can be stored, shared and analysed with the help of computers; services have also been developed to help distribute the data and to advise researchers on how to use it. Unsurprisingly, the use of census data has expanded, while administrative data is still used on a more occasional basis. It has been suggested that more use could be made of the latter data in areas such as nursing research (see Herron, 1989; Jacobson et al., 1993; Lobo, 1986; Reed, 1992; Woods, 1988).[4]

Since the 1960s, various types of social surveys have been conducted by government agencies, academics and commercial organizations and the resulting data made available for secondary research purposes. Hakim (1982) distinguishes multi-purpose (or omnibus) social surveys from those designed with an exclusive primary focus. As the name suggests, multi-purpose surveys are carried out in order to provide data for multiple uses (and users). In the United Kingdom, examples of multi-purpose surveys include the General Household Survey (GHS) and Labour Force Survey (LFS), which are carried out by the Office for National Statistics (ONS). The British Social Attitudes Survey (BSA) conducted by the National Centre for Social Research (NatCen) is another example of a survey carried out to provide data for others to analyze. Its American equivalent, the General Social Survey (GSS), has been carried out annually since 1971 by the National Opinion Research Center (NORC) based at the University of Chicago. Cross-national projects have also been developed, such as the International Social Survey Programme (ISSP) (Procter, 1995).

In contrast, more exclusive social surveys are designed to investigate particular research questions and are conducted on a one-off or less regular basis. Given that such surveys are designed to examine specific issues, the scope for secondary analysis may be more restricted compared to data derived from omnibus social surveys (Dale et al., 1988: 9). The same caveat applies to other statistical data collected for specific rather than generic research purposes. However, these data sets are still used as a resource for secondary research purposes. For example, data from the Nuffield study of social mobility in England and Wales (Goldthorpe, 1980) have been re-used on various occasions.

Some statistical data sets have been generated with a view to both primary and secondary research uses. For instance, longitudinal studies follow up a population cohort over time, collecting data on topics of long-term interest and further topics for investigation may be introduced as the study progresses. These data are subject to primary analysis by the researchers who collected the data to address particular research questions; they may also be archived and used by other researchers to address additional research questions. Longitudinal studies have been most extensively carried out in the United States. The oldest ongoing American longitudinal project is the Lewis Terman study which began in 1921/2. It was originally designed to investigate the maintenance of early intellectual superiority over a ten-year period, and subsequently expanded to examine the life paths into adulthood of gifted individuals and the experiences of these individuals and their families. Data from the Terman database have been used in secondary studies, for example, Elder et al. (1993) used its data on World War II veterans in their secondary study of military service in World War II and its effect on adult development and aging.

What is Secondary Analysis?

Table 1.2 Examples of pre-existing qualitative data used in social research

Type	Examples	Methodology
Non-naturalistic or artefactual data (solicited for research studies)	Fieldnotes Observational records Interviews Focus groups Questionnaires (responses to open-ended questions) Diaries (solicited) Life stories	Secondary analysis
Naturalistic data (found or collected with minimal interference by researchers)	Life stories Autobiographies Diaries (found) Letters Official documents Photographs Film	Documentary analysis
	Social interaction	Conversation analysis

Longitudinal studies have also been carried out in other countries, though not to the same extent. Hakim (1982) describes five national longitudinal studies carried out in England and Wales or Great Britain, of which only three followed up the sample beyond two years: the OPCS Longitudinal Study (LS), the National Child Development Study (NCDS) and the National Survey of Health and Development (NSHD). Other non-national longitudinal projects carried out in England include the Newsons' study of child development (Newson and Newson, 1963, 1968, 1975). New 'linked' data sets may also be generated by combining selected data from these resources (Dale et al., 1988). For example, the aforementioned OPCS Longitudinal Study, now known as the ONS Longitudinal Study (LS), links data obtained from the 1971 Census onwards with data obtained annually from public records for those born on one of four days of the year.

Qualitative data

In contrast to quantitative research where, as we have seen, secondary analysis encompasses the use of pre-existing data derived from research and other contexts, in qualitative research different methodologies have been used for the analysis of 'non-naturalistic' data that were solicited by researchers, and 'naturalistic' data that were 'found' or collected with minimal structuring by researchers (see Table 1.2).[5,6]

In qualitative research secondary analysis is more narrowly conceptualized as a methodology for the study of non-naturalistic or artefactual data derived from previous studies, such as fieldnotes, observational records, and tapes and transcripts of interviews and focus groups. However, unlike quantitative research, there is no tradition of re-using data from previous qualitative studies. On the occasions where research data have been re-used it has tended to be in exceptional circumstances, such as when anthropologists have analysed the fieldnotes of researchers who have died before completing their work and published the results posthumously (Sanjek, 1990a). It is only relatively recently that researchers have begun to recognize the potential for re-using the various types of qualitative data produced in the course of social research, and to publish secondary studies using these resources (Heaton, 1998).

By contrast, naturalistic data such as diaries, essays and notes, autobiographies, dreams and self-observation, photographs and film have traditionally been studied using the more established methodology of documentary analysis (Jupp and Norris, 1993; Plummer, 1983, 2001; Scott, 1990). Documentary analysis has also been used to study some types of textual and non-textual data from qualitative studies which are near-naturalistic in that they were obtained with minimal shaping by researchers (as in unstructured interviews), or by using unobtrusive or even covert methods. However, some types of qualitative data can be construed as naturalistic or non-naturalistic, depending on how they originated. For example, diaries can be kept as a personal record and later 'found' and examined by researchers; alternatively, they can be designed and completed as a research tool. Similarly, life stories may be told and recorded in autobiographies, biographies or interviews, and may be more or less structured by journalists, biographers or researchers. Thus, although secondary analysis and documentary analysis tend to focus on non-naturalistic and naturalistic data respectively, this distinction is not always clear-cut and hence there is some overlap in the types of data which are subject to these methodologies.

Life stories solicited for qualitative studies are unique data in that although they tend to be collected primarily for single use, as part of a research study investigating a specific question, they are also recorded with the intention of archiving them for possible future use in other research. For this reason they are similar to the kind of statistical data obtained in longitudinal studies in terms of being collected with both primary and secondary uses (and users) in mind. However, although life stories have been recognized as potentially valuable secondary resources (Atkinson, 1998; Thompson, 1998) and probably archived more than any other varieties of qualitative data, it is not clear to what extent, and how,

What is Secondary Analysis?

researchers have used these pre-existing data. There is little published national or international information on the extent and nature of use of these resources generally.[7] In addition, there have been no recent reviews of the use of life stories in social research and whether these were obtained from archives or through primary research.[8] There is also little guidance on how to do social research using pre-existing life stories: textbooks tend to focus on how to collect, record and analyze life stories in the context of primary research rather than how to access and undertake secondary research using pre-existing life stories deposited in archives (see Atkinson, 1998; Langness, 1965; Miller, 2000; Yow, 1994).

The nearest qualitative equivalent to omnibus social surveys designed to provide statistical data for use in multiple studies are projects which have been undertaken to collect and preserve textual and non-textual remnants of social life. The Mass Observation project is a prime example of this type of work. Described as an 'anthropology of ourselves' (quoted in Calder and Sheridan, 1984: 247), various materials were gathered by the investigators on the project between the 1930s and 1950s and more recently during the 1980s and 1990s. Various non-naturalistic and naturalistic data were collected, including reports of interviews and observations, time sheets and diaries, and original documents, such as press cuttings, pamphlets, and photographs. These data were collected and archived with a view to their value as a historical resource and more general utility. The archivist (Sheridan, 2000) notes that these data have been used by academic researchers from a range of disciplines (including art history, social history, anthropology, psychology, sociology, media and cultural studies) as well as by the wider community (including the media, novelists, artists, teachers and students). She adds that since 1970 social research using the data has taken four main forms: as evidence in historical research; in the study of the Mass Observation movement itself; in methodological research; and to inform the development of related initiatives. In principle, the re-use of some of the data from the Mass Observation project may therefore be defined as documentary or secondary analysis, depending on which type of data is used and for what purposes (of which more below).

Another methodology which has been used in social research to study naturalistic data in the form of recordings of everyday social interaction is conversation analysis (see Atkinson and Heritage, 1984; Hutchby and Woofit, 1998; ten Have, 1999). These data include tapes and transcripts of audio and/or visual recordings made and transcribed by the researchers personally, or by other researchers. Indeed, conversation analysts often develop analyses based on pre-existing data that have been shared within this disciplinary network. Lengthy extracts of the data, transcribed using a detailed system of annotating verbal and non-verbal interaction, are also reproduced in published studies in order to allow for independent access to, and scrutiny of, the data upon which the analyses are based. However, despite these practices, no distinction is made between 'primary' and 'secondary' conversation analysis. Rather, these naturalistic data are seen as pristine and in no way an artefact of the research process; as such, they are assumed to be open to analysis by all on an equal basis, regardless of who collected the data and rendered it for analysis. Conversation analysis is therefore an interesting example of a methodology that involves the re-use of qualitative data but which is not regarded by exponents as a secondary methodology because of the perceived unadulterated and naturalistic nature of the data concerned.

Functions of secondary analysis

It has been suggested that quantitative and qualitative secondary analysis can be used for three broad purposes although, as I show below, some of these claims are a matter of debate.

Investigation of new or additional research questions

First, it has been proposed that secondary analysis allows researchers to put to new or additional uses data that were originally collected for other research purposes or, in the case of quantitative data, administrative purposes. This is illustrated by the following quotations, in which the secondary analysis of quantitative data is defined as:

> ...the study of specific problems through analysis of existing data which were originally collected for other purposes. (Lipset and Bendix, 1959. Quoted in Glaser, 1962: 71)

> ...the *extraction of knowledge on topics other than those which were the focus of original surveys.* (Hyman, 1972: 1. Original emphasis)

> ...any further analysis of an existing data set which presents interpretations, conclusions, or knowledge additional to, or different from, those presented in the first report on the inquiry as a whole and its main results. (Hakim, 1982: 1)

> ...[the study of] a problem by analysing data that originally were collected for another study with a different purpose. (Woods, 1988: 334)

Likewise, it has been suggested that the secondary analysis of qualitative data involves:

the use of an existing data set to find answers to a research question that differs from the question asked in the original or primary study. (Hinds et al., 1997: 408)

Thus, quantitative and qualitative secondary analysis have each been conceptualized as methodologies for conducting free-standing studies using pre-existing data originally collected for other purposes.

Although this proposition is generally accepted, it has been pointed out that this way of defining secondary analysis is limited in that it fails to acknowledge that some projects are actually designed to supply data for sundry secondary studies (Dale et al., 1988). The data from omnibus surveys are, for example, collected and prepared for secondary research purposes and are not subject to primary analysis as such. To a lesser extent, the same applies to data from longitudinal studies which are designed both to address particular primary research questions and to supply long-term data for future secondary research.

Verification, refutation and refinement of existing research

Second, and more controversially, it has been suggested that secondary analysis can be used as a means of verifying, refuting or refining the findings of primary studies through the re-analysis of data sets. This proposition is illustrated by the following quotations, where it is suggested that the secondary analysis of quantitative data involves the:

>...re-analysis of data for the purpose of answering the original research question with better statistical techniques, or answering new questions with old data. (Glass, 1976: 3)
>
>...re-analysis of data originally collected and analysed by another investigator addressing the same question, a different question, or applying different methods of analysis. (Woods, 1988: 334)
>
>...use of data gathered for the purposes of a primary research analysis (original research) but looks at questions not addressed by the original investigator, or addresses the same questions using different methods of analysis. In large-scale studies, secondary analysis is further used by investigators to validate the results from the primary analysis; that is, re-analysis and testing of new hypotheses may support further or dispel initial findings. (McArt and McDougal, 1985: 54)

This principle is generally accepted in quantitative secondary analysis, although it too has been queried. For example, Hyman (1972: 76, footnote 3) argues that the re-analysis of survey data for the purpose of verification is not a type of secondary analysis on the grounds that 'the focus [of re-analysis] is on the original problem that the survey was intended to illuminate' and not on a new topic. It could be countered, however, that re-analysis nevertheless has the potential to generate additional knowledge and insights through the production of findings which may or may not corroborate the primary work. In addition, where such studies are carried out by independent researchers, re-analyses are also secondary to, and free-standing of, the original research in the sense of being carried out by other analysts, even though each study investigates the same questions using the same data.[9]

To a lesser extent, secondary analysis has also been suggested as a possible means of re-investigating questions addressed in previous qualitative studies (see Corti, 2000; Szabo and Strang, 1997). For example, Szabo and Strang (1997: 67) have claimed that it can involve looking at 'the same questions with different analysis methods'. However, as Hammersley (1997a) has pointed out, whether the findings of qualitative research can be verified in the same ways as quantitative research is debatable, given that there are different philosophical perspectives on the respective nature of quantitative and qualitative data and associated research methods.[10] Qualidata have also identified resistance to the re-use of qualitative data for this purpose (Corti, 1999; Corti et al., 1995). This proposition appears therefore to be more contentious in relation to qualitative secondary analysis, reflecting wider epistemological tensions in quantitative and qualitative research which will be explored in later chapters.

Synthesis of research

The last and most contentious of the purported functions of secondary analysis is whether or not it encompasses various types of meta-research designed to synthesize knowledge arising from existing studies. Meta-analysis and related techniques for examining evidence from quantitative research have been described by some commentators as forms of secondary analysis (Jensen and Allen, 1996; Woods, 1988). However, others have stressed that these various meta-research strategies are distinct from secondary analysis in that they seek to identify, appraise, and aggregate or synthesize existing knowledge on a particular topic. Thus, Glass distinguishes meta-analysis from secondary analysis, describing the former as the 'analysis of analyses' (1976: 3). Kielcolt and Nathan also claim that meta-analysis is unique in that it seeks to integrate the findings from a 'universe' or 'sample' of investigations of some phenomenon: 'that is, the study itself becomes the unit of analysis' (1985: 10).

Similarly, some commentators have claimed that qualitative versions of meta-analysis are forms of secondary analysis. For example, 'aggregated analysis', which involves the synthesis of findings from several qualitative studies, has been described as a variety of secondary analysis

(Estabrooks et al., 1994). However, in contrast, Thorne has argued that 'whereas meta-analysis serves as a strategy for synthesis of research findings, secondary analysis provides a mechanism for extending the contexts in which we are able to use and interpret qualitative research data' (1998: 548). In addition, I have previously suggested that meta-analysis, meta-synthesis, meta-studies, systematic reviews, narrative analysis and literature reviews differ from secondary analysis on two grounds. First, the former research strategies are concerned with appraising and summarizing existing knowledge, and not with exploring new research questions or verifying the results of individual studies. Secondly, they mainly involve the study of research reports, seldom reverting to the raw data itself (Heaton, 1998). For these reasons, I regard quantitative and qualitative meta-research strategies as separate methodologies from secondary analysis.[11]

However, some forms of meta-research are more difficult to distinguish from secondary analysis than others. For example, in quantitative research, there is an approach involving so-called 'individual' meta-analysis in which researchers do revert to the raw data in order to standardize analyses across studies and/or to re-examine the data in new ways (Smith and Egger, 1998). This may be necessary for the purposes of synthesizing results when they are not reported in a standard way, or cannot be dis-aggregated, allowing findings relating to particular groups to be scrutinized. While this procedure is consistent with the meta-analyst's aim of synthesizing existing findings, the researcher is also, in effect, adding to what was previously known through the re-modelling and re-analysis of the micro data. In producing new knowledge, the boundaries of this form of meta-analysis therefore blur with those of the secondary analysis of data sets.[12]

Similarly, in qualitative research, Noblit and Hare contend that meta-ethnography: 'can be considered a complete study in itself. It compares and analyzes texts, creating new interpretations in the process. It is much more than we usually mean by a literature review' (Noblit and Hare, 1988: 9). In 'creating new interpretations' this approach appears to overlap with secondary analysis. Conversely, it is debatable whether 'aggregated analysis' is, as the authors (Estabrooks et al., 1994) claim, a variety of secondary analysis or in fact a meta-research strategy. On the one hand, theory development is the primary aim of aggregated analysis, not merely the synthesis and comparison of findings; in this respect it resembles secondary analysis. On the other hand, in aggregated analysis the researcher does not return to the original data. Instead, the findings reported by the original researcher are used – although the authors note that, unlike meta-ethnography (and presumably other meta-research strategies), aggregated analysis involves 'interpretative techniques' and is not a 'context-stripping

Table 1.3 Modes of secondary analysis

Mode	Source of data	Data donors
Formal data sharing (via intermediary services)	General data archives Special collections Commercial companies Public records Publications	Primary researchers for use in independent secondary studies Organizations for use in independent secondary studies
Informal data sharing (by special request and private networks)	Primary researchers' personal data collections Disciplinary networks Organizations' in-house records	Primary researchers for use in secondary studies by independent researchers Primary researchers for use in secondary studies by independent researchers *and* the data donors Organizations for non in-house secondary use
Personal or inside secondary analysis	Primary researchers' personal collections Organizations' in-house records	Primary researchers for use in personal secondary analyses of their own data Organizations for secondary use in-house

activity' (Estabrooks et al., 1994: 505). Thus, aggregated analysis appears to be a hybrid methodology, drawing on aspects of secondary analysis and meta-analysis in its approach to social research using pre-existing data.

Modes of secondary analysis

There are three main modes of secondary analysis (see Table 1.3). In the first – 'formal' data sharing – secondary analysis is carried out using data sets that have been officially made available for data sharing (although access may be controlled or restricted). This includes data sets deposited in general data archives and special collections, and those managed by commercial companies, as well as raw data published in research reports or other media. In this mode of secondary analysis, the data will have been collected and deposited by one group of researchers or organizations (data donors) and accessed by another (data users), hence secondary studies using these resources are carried out by independent researchers.

This is not necessarily the case in 'informal' data sharing where data is either obtained directly from primary researchers and organizations by request, or indirectly through private disciplinary networks (as in conversation analysis). In this mode of data sharing, the secondary analysts may or may not invite the primary researchers who donated the data to be part of the research team carrying out the secondary research. Two or more primary researchers may also combine or pool data sets from their previous work and jointly analyze them as part of a new secondary project; in this case, as primary-cum-secondary analysts, the researchers will be working with a mix of independently and self-collected data sets.

In contrast, the last mode of secondary analysis does not involve any data sharing. Instead, researchers re-use their own data, or what I referred to as 'auto-data' in Heaton (2000), and organizations carry out internal secondary analyses of their own records. Personal or inside secondary analysis is unique in that it is carried out by the same researchers and organizations that originally compiled the data, and no one else.

In existing conceptualizations of the secondary analysis of quantitative data it is often assumed that such studies are carried out using data which have been collected by other researchers,[13] although whether this is via formal or informal data sharing is not usually specified. This is illustrated by several of the above quotations, as well as the following, in which the methodology is described as:

> ... any analysis of data collected originally by persons other than the analyst. (Miller, 1982: 719)
>
> ... a form of research in which the data collected by one researcher are reanalysed by another investigator, usually to test new hypotheses. (Polit and Hungler, 1983. Quoted in McArt and McDougal, 1985: 54)
>
> ... the application of creative analytical techniques to data that have been amassed by others. (Kiecolt and Nathan, 1985: 10)
>
> ... the further analysis of data by anyone other than those responsible for its original commissioning or collection. (Dale et al.,1988: 4)

However, Hyman (1972) has shown that the secondary analysis of quantitative data has not always been the prerogative of independent analysts. In discussing the secondary analysis of survey data he distinguishes pure from what he calls 'semisecondary' analysis. While the former is exemplified by analysts who make use of survey data deposited in data archives, the latter refers to three other types of secondary analysis, one of which is performed by the original, primary investigators. According to Hyman, this approach was in practice before the era of data banks, when researchers drew upon the wealth of data that they or their organizations had amassed themselves.

While the re-use of auto-data has been largely disregarded in the more recent literature on the secondary analysis of quantitative data as attention has shifted to the use of major, independently collected data sets, this and other modes of secondary analysis have all been recognized in work on the re-use of qualitative data sets. For instance, Thorne (1994, 1998) envisaged at least five discrete types of qualitative secondary analysis, involving both the re-use of other researchers' data sets and researchers' own data from their previous primary work. In the first variety, called 'analytic expansion', researchers make further use of their own data 'to answer new or extended questions' (Thorne, 1998: 548). In 'retrospective interpretation', researchers examine new questions which were raised but not addressed in the context of the primary study (although here Thorne does not specify whether this applies to auto-data sets and/or independently collected data). The third type of secondary analysis is used in relation to other researchers' data sets, which are subject to 'armchair induction'. Using this approach, researchers apply inductive methods of textual analysis for the purposes of theory development. A fourth and additional approach to theory development using secondary analysis is that of 'amplified sampling'. This involves the comparison of several distinct and theoretically representative data sets. The fifth and final approach outlined by Thorne is 'cross-validation'. Again this involves the use of pre-existing, independently collected data sets, this time in order to 'confirm or discount new findings and suggest patterns beyond the scope of the sample in which the researcher personally has been immersed' (Thorne, 1994: 267).

Other authors have also examined the potential of doing qualitative secondary analysis using these different sources of qualitative data. For example, Szabo and Strang (1997) and Mauthner, et al. (1998) discuss their respective experiences of re-using qualitative data that they personally had previously collected and the issues raised by this mode of secondary analysis. In addition, West and Oldfather (1995) have outlined a methodology called 'pooled case comparison' which is based on the informal sharing of qualitative data. Defined as a way of 'comparing separate but similar studies *ex post facto*', it involves the use of raw data from two primary studies while setting aside any 'categories and properties from previous analyses' (West and Oldfather, 1995: 454). The approach is similar to the 'amplified sampling' form of secondary analysis outlined by Thorne (see above), in which comparisons across two or more independently collected data sets are made, although just two data sets are used, each of which were originally collected by one or other of the secondary analysts. Finally, Qualidata and other archivists have highlighted the potential of re-using qualitative data sets which have been formally archived and made available for data sharing (see Corti et al., 1995; Corti and Thompson, 1998; James and Sørenson, 2000).

Conclusion

'Secondary analysis' is a rather nebulous concept. It has been variously associated with different types of quantitative and qualitative data, different functions, and different *modi operandi*. Conceptually, there are similarities in the ways in which quantitative and qualitative secondary analysis have been defined, and some overlap between these approaches and other methodologies, such as meta-analysis and documentary analysis. However, there are also some notable variations in these concepts which suggest that quantitative and qualitative secondary analysis are not equivalent research strategies, and that these methodologies can be broadly distinguished from related research strategies.

Three key variations in the conceptualizations of quantitative and qualitative secondary analysis have been observed. The first concerns the types of pre-existing data which are deemed to be subject to secondary analysis. While quantitative secondary analysis encompasses the re-use of data derived from research and other contexts, qualitative secondary analysis has a narrower focus on non-naturalistic data derived from previous qualitative studies. Second, while it is generally accepted that quantitative and qualitative secondary analysis can be used to investigate new or additional research questions, it is a matter of contention as to whether secondary analysis can be used to verify previous qualitative research in the same ways as it is in quantitative research. Finally, another difference in quantitative and qualitative secondary analysis concerns the sources of data which may be used. In quantitative secondary analysis, it is generally assumed that it involves the use of other researchers' data, obtained via formal or informal data sharing. However, in qualitative secondary analysis, the potential of re-using one's own data has also been recognized alongside these modes of secondary analysis.

The focus of qualitative secondary analysis on non-naturalistic data derived from previous studies distinguishes it from other qualitative methodologies, such as documentary analysis and conversation analysis, that are used to study more naturalistic pre-existing qualitative data which are not an artefact of research work. However, as we have seen, some qualitative data, such as life stories, can be construed as naturalistic or non-naturalistic depending on how they originated and hence there is, in principle, some overlap between the methodologies used to examine these types of data. Similarly, while there is in principle some overlap between secondary analysis and quantitative and qualitative varieties of meta-analysis, I have suggested that they are broadly distinguishable in terms of the functions of these approaches and the nature of the pre-existing material that is involved. Thus, secondary analysis is not regarded as a methodology for the synthesis of previous research, but as a methodology for investigating new research questions or for verifying previous research. However, it is recognized that the latter function is also a matter of contention in relation to qualitative secondary analysis in particular.

These findings raise issues about whether existing conceptualizations of quantitative and qualitative secondary analysis are reflected in practice. For example, it is one thing to claim that secondary analysis may be used to verify research but it is another as to whether it has been used for this purpose in the qualitative secondary studies conducted to date.[14] Similarly, it would be interesting to know to what extent researchers have used their own or other researchers' qualitative data in the secondary studies carried out to date. In order to further investigate the ways in which secondary analysis has been developed and actually used as a qualitative methodology to date, the following, provisional, definition of secondary analysis was adopted as the basis for an empirical review of existing qualitative secondary studies:

> *Secondary analysis is a research strategy which makes use of pre-existing quantitative data or pre-existing qualitative research data for the purposes of investigating new questions or verifying previous studies.*

This formulation qualifies the type of qualitative data involved (limited to data from research studies) and recognizes two of the proposed functions of secondary analysis (but not the synthesis of research findings). At the same time, it leaves open the issues of whether or not qualitative secondary analysis can be used to verify the results of previous work, and which sources of qualitative data are used. It should be stressed that this is a working definition, the adequacy of which will be examined throughout the book, in the light of the findings of the review of existing qualitative secondary studies. Before I describe the studies that were identified, in the next chapter I first of all examine the context in which qualitative secondary analysis emerged and became, to borrow Denzin and Lincoln's (1998) phrase, part of the 'landscape' of qualitative research.

Notes

1. Some of this literature has focused on the potential of re-using statistical data in particular areas of social research, such as health (see Adams et al., 1994; Gleit and Graham, 1989; Gooding, 1988; Herron, 1989; Jacobson et al., 1993; McArt and McDougal, 1985; Woods, 1988), education (see Burstein, 1978; Glass, 1976; Miller, 1982), and criminology (see Riedel, 2000).

2. While this view is generally accepted, Woods (1988) suggests that the initial analysis of, say, clinical data for research purposes is in fact primary analysis. She

therefore retains secondary analysis as a term applying only to data analysed in the course of a prior study.

3. These resources have been described in more detail elsewhere and are simply summarized here (see Dale et al., 1988; Elder et al., 1993; Hakim, 1982; Hyman, 1972).

4. For example, Lobo observes that the following administrative data can be subject to secondary analysis:

> Nurses collect or generate an immense amount of data during research and in clinical practice. Many times these data may or may not have been analysed during the initial evaluation of the data. These data may or may not have been organized in a systematic manner when they were initially collected. Clinical data available to nurses for research include nursing notes, vital sign sheets, and input and output sheets. The documents created as a result of nursing research and nursing practice can provide an immense amount of information about diagnosis and treatment of human responses to actual or potential health problems. These documents can be used as resources and can provide information about what kinds of actions nurses take and the impact of those actions on their individual clients. The research strategy used to analyze material for purposes other than that for which it was originally collected is termed secondary analysis. Although this strategy is usually considered when data collected specifically for research purposes are available, as nurses, we also have access to a large amount of data that are naturally generated as a result of our practice. (1986: 295)

5. The distinction between 'quantitative' and 'qualitative' data, and 'naturalistic' and 'non-naturalistic' data, is not meant to suggest that the re-use of these materials is necessarily segregated in social research. On the contrary, qualitative data may be collected in what are primarily quantitative studies and subject to secondary analysis, as in the case of open-ended responses to survey questionnaires; qualitative data may be subject to more quantitative forms of content analysis. In addition, naturalistic and non-naturalistic data may be combined and jointly analysed in qualitative research studies.

6. The term 'naturalistic' is here used after Plummer (2001) who uses it to distinguish three broad types of life stories: 'naturalistic life stories' which are told as part of everyday life; 'researched and solicited stories' which are mediated by researchers; and 'reflexive and recursive life stories' which are concerned with the telling of the story as well as the story itself.

7. Corti and Thompson (1998) report that interviews from the latter's study, 'Family life and work experience before 1918', archived at the University of Essex since the 1970s, have been seen by over one hundred researchers and students and re-used in numerous books and articles. Atkinson reports that the use of life stories deposited with the Centre for the Study of Lives at the University of Southern Maine in the United States from 1988 was 'slow to develop' (Atkinson, 1998: 4).

8. Denzin (1970) reports that a review of 22 studies that used the life history method between 1920 and 1940 was carried out by Angell (1945). Most of the classic studies using life stories appear to have been based on those collected for primary research purposes rather than those from pre-existing collections. For example, one of the best known studies using a life story is Thomas and Znaniecki's (1958) *The Polish Peasant in Europe and America*. One of its original five

volumes (published 1918–20) comprises a 300 page life story of Wladek Wisniewski written shortly before the outbreak of World War I, covering his early life in Poland through to his experiences as a Polish immigrant living in America. His life story was solicited for the study and analysed together with documents in the form of letters, newspaper articles and official public records. Strauss and Glaser (1977) also gathered Mrs Abell's life story for their study: *Anguish: A Case History of a Dying Trajectory*.

9. This is distinct from replication or re-studies in which new data *are* collected in order to re-investigate the questions examined in previous work (see Chapter 3 for more on this topic).

10. For a useful further discussion of related epistemological issues, see Bryman (1988), Hammersley (1997b) and Seale (1999).

11. It may be added that there is a growing literature specifically on techniques for synthesizing the results of quantitative and qualitative research in medicine and, increasingly, in other areas of social research (see Dixon-Woods et al., 2001; Jensen and Allen, 1996; Paterson et al., 2001; Popay et al., 1998).

12. While the possibility of doing meta-analysis using raw data has been recognized for some time (see Woods, 1988: 335), it is still not the norm in meta-research.

13. For an exception, see Herron who, writing on the secondary analysis of quantitative data, has suggested that these data 'may have been gathered earlier and then re-examined by the same researcher...' (1989: 66).

14. While it is beyond the scope of this book to examine secondary studies of quantitative data, it would be interesting to update Hyman's (1972) review of the nature and use of quantitative secondary analysis. In particular, it would be useful to establish the extent to which re-analyses (and re-studies) of research based on quantitative data have been conducted in practice in the social, natural and biomedical sciences.

2

From Quantitative to Qualitative Secondary Analysis

Developments in data archiving and computing	20
The promotion of data sharing and data retention	22
Potential benefits and problems of re-using data	26
Conclusion	31
Notes	32

> The history of secondary analysis may have curbed as well as stimulated its use, since all writers to date have focused on the importance of secondary analysis of survey data ... the emphasis on survey data neglects other kinds of data, particularly field data, and hence limits the potential use of secondary analysis. This research strategy can be applied to almost any qualitative data however small its amount and whatever the degree of prior analysis. Survey data may always remain the primary source of material for secondary analysis. However, with the current increase in all types of data collection, it is by no means the only possible source. Secondary analysis is something that the sociologist can do with data of his own choosing. (Glaser, 1963: 11)

Despite Glaser's advocacy, survey data has remained the main source of material used in secondary studies carried out by social researchers. It is only recently that researchers and archivists in North America and Europe have begun to investigate and debate the possibility of doing secondary research using qualitative data (see Corti, 1999, 2000; Corti et al.,1995; Corti and Thompson, 1998; Fink, 2000; Hammersley, 1997a; Heaton, 1998; Hinds et al., 1997; Humphrey et al., 2000; James and Sørensen, 2000; Kuula, 2000; Leh, 2000; Mauthner et al., 1998; Szabo and Strang, 1997; Thorne, 1994, 1998). In this chapter, I examine how the delayed development of qualitative secondary analysis has been shaped by the following factors: the emergence of data archives and advances in computing; the promotion of data sharing and data retention in policy and practice; and the varied views of academics on the potential benefits and problems of data sharing in the social sciences.

Developments in data archiving and computing

Prior to the advent of data archives, the main sources of secondary data were public records, organizations' in-house records, publications, and researchers' personal data collections. Since the mid-twentieth century the availability of secondary data has greatly increased as a result of developments in data archiving and technological innovations in computing. Quantitative data sets, in particular, are now widely and readily available in electronic form from data archives. While these developments in data archiving and advances in computing have been mainly applied to quantitative data sets, as I show below, in recent years they have been extended to qualitative data sets, leading to an increase in the availability of these resources.

One of the first major archiving projects in the social sciences began in 1937 with the creation, by George Peter Murdock and colleagues, of the Human Relations Area Files (HRAF) at Yale University in the United States. The aim of the project, which is ongoing today, was to catalogue the results of anthropological research and to make these available for secondary historical and cross-cultural research.[1] Since its beginning the emphasis of the HRAF has been on coding deposits for statistical analysis. As Turner (1985) observes, this accent shows the influence on Murdock and the project of Herbert Spencer's work. Spencer (1873–1881) attempted to catalogue the characteristics of societies in 15 volumes of *Descriptive Sociology*, the first of which was published in 1873, drawing on biologists' work on the classification of species. In following this tradition, the HRAF project has been described as a 'positivistic project par excellence' in cultural anthropology (Wax, 1997: 19).[2] More generally, the subsequent tendency of archives to focus mainly on quantitative data sets reflects the continuing influence of the positivist paradigm in the social sciences.

The first general data archive (or data bank) was founded in the United States in 1957 when the in-house archive of The Roper Center for Public Opinion Research became a depository for data sets from various sources. Other archives were subsequently established in the United States and elsewhere, set up as national resources or as special collections held by Universities and other establishments. Early examples include the Zentralarchiv für empirische Sozialforschung in Germany which was created in 1960, the Social Science Research Council Survey Archive in the

United Kingdom which was set up in 1967 (becoming the UK Data Archive in 1984), and the Danish Data Archives (DDA) which were founded in 1973. General data archives have tended to deal mainly in statistical data arising from social surveys, longitudinal studies and censuses. However, a minority have also accepted data from qualitative studies. For example, the Murray Research Center: A Center for the Study of Lives at the Radcliffe Institute for Advanced Study at Harvard University in the United States holds over 270 data sets from research on human development and social change, including longitudinal data sets with a qualitative component (James and Sørensen, 2000).[3] Some qualitative studies are also deposited with the UK Data Archive and with the National Social Policy and Social Change Archive at the University of Essex in England. Elsewhere in Europe, the Finnish Social Science Data Archive (FSD) accepts qualitative data (Kuula, 2000) and the Danish Data Archives (DDA) have recently expanded their intake to include qualitative data sets (Fink, 2000).[4]

In addition to these general data archives, some specialist archives for qualitative data, focusing mainly on naturalistic material, have also been established. There are a number of archives for life and oral histories and related biographical materials in the United States and Europe. These include The Centre for the Study of Lives at the University of Southern Maine in the United States which was set up in 1988 (Atkinson, 1998), The Verein zur Forderung von Forschungen zur politischen Sozialisation und Partizipation (POSOPA) e.V. in Germany which holds biographical interviews from research projects on youth in the twentieth century (Groeschel, 2000), and the archive of the Institut für Geschichte und Biographie in Germany which holds oral history interviews on 'Deutsches Gedächtnis' (German memory) (Leh, 2000). Another notable archive that specializes in the archiving of more naturalistic qualitative materials is the Tom Harrison Mass-Observation Archive which was founded at the University of Sussex in England in 1975. It contains the records of the full-time Mass Observation investigators and the contributions of the volunteer observers, including day surveys and personal diaries (Calder and Sheridan, 1984).[5]

In an effort to promote the archiving of all types of qualitative data across the social sciences, the Economic and Social Research Council (ESRC) established the Qualitative Data Archival Resource Centre (Qualidata) at the University of Essex in England in 1994; this became an integral part of the UK Data Archive in October 2001.[6] Although not an archive itself, Qualidata facilitates the archiving of qualitative data in existing depositories throughout the United Kingdom (Corti et al., 1995). When it was established, Qualidata estimated that only around ten per cent of qualitative studies were being archived (Corti, 1999). Since then, the number of qualitative data sets catalogued in archives in the United Kingdom has risen from approximately 100 in 1995 to 460 in 2000 (Corti, 2000).[7] Information on the availability and form of these data sets has also been improved by the creation of Qualicat, a searchable online catalogue of these resources.[8]

While the methods for archiving and sharing quantitative data sets have been transformed by advances in computing, qualitative data such as transcripts and fieldnotes are still generally preserved in their original format and accessed *in situ* rather than distributed via electronic means. However, ongoing technological innovations are very likely to transform current practices in the archiving and sharing of qualitative data sets in the near future. These include the digitization of qualitative data, the growing use of computer-assisted qualitative data analysis software (CAQDAS), the development of software for facilitating the processing and exchange of qualitative data sets, and the electronic archiving of data sets (see Corti, 1999; Fielding, 2000; Muhr, 2000). Some data sets can already be accessed through the Internet, such as Stirling's anthropological work on two Turkish villages (Zeitlyn, 2000). Thus, as Fielding (2000: para 25) has observed, the 'dream scenario' of a web site providing access to qualitative data sets around the world that can be downloaded for secondary analysis is not an 'impossible dream': the technology for this does exist at present but it has not been generally applied as yet.

The promotion of data sharing and data retention

The establishment of data archives has been supported by various official organizations and groups which have encouraged social researchers to retain and share their data on both an informal and formal basis. As I show below, the rise of a data sharing imperative in the social sciences began in the United States in the 1960s and has focused mainly on research involving quantitative rather than qualitative methods, although there is growing policy and academic support for the latter.

The first major official report on data sharing in the social sciences was published in 1985 by the National Academy of Sciences (NAS) in the United States, following a conference in 1979 and subsequent exploration of the matters raised there by a panel of the Committee on National Statistics (Fienberg et al., 1985). Focusing exclusively on issues relating to the sharing of quantitative data, the NAS report concluded that the sharing of this type of data was in the best interests of open scientific inquiry and should be encouraged where feasible. The panel also produced sixteen recommendations for various groups, designed to promote both informal and formal data sharing (see Table 2.1). Emphasis is placed on data sharing as a means of enabling research claims to be independently verified, refuted or refined, as well as a way of safeguarding against fraudulent claims, particularly in research influencing public policies.[9] The panel also suggested that, in order to allow research to be independently examined, the 'availability of the data for scrutiny and re-analysis should be part of the presentation of results' and lamented the decline of this practice, citing Cavendish's 1798 paper on the density of the earth as a 'prime' example of this tradition (Fienberg et al., 1985: 9).[10] It also expressed support for formal data sharing by encouraging sponsors of social research to address the need for more archives for general purpose data sets.

Some but not all of the panel's recommendations have been heeded by the groups concerned. For example, some journals in the social sciences have developed guidelines recommending that authors keep their data for at least five years in case the findings are queried and need to be verified (Sieber, 1991). However, in general, support for data retention is more common in areas of social science which rely on quantitative rather than qualitative methods, and in the natural and biomedical sciences where some sponsors of research also define this as good practice.[11]

Formal data sharing has also been promoted by some funders of social research as a means of gaining additional value from previous studies by enabling the data to be used for multiple purposes where possible. This has been the policy of sponsors, such as the Economic and Social Research Council, which have funded the establishment of data archives in order to make data sets more generally available for further use. The Economic and Social Research Council also requires applicants for funding to consider whether or not their proposed research could be carried out using existing data sources. If not, those proposing to do primary research are asked whether or not they would consider depositing their data set in an archive for possible secondary analysis by others (and, if so, they can request additional funding to cover the costs of processing the data for archiving). These policies were initially applied to applicants planning quantitative studies and have since been extended to those submitting qualitative designs. Other sponsors that encourage primary researchers to share and/or carry out secondary research include The Wellcome Institute, Joseph Rowntree Foundation (JRF) and Nuffield Foundation in the United Kingdom; the National Science Foundation (NSF), National Institutes of Health (NIH) and National Institute of Justice (NIJ) in the United States; and the Social Sciences and Humanities Research Council (SSHRC) in Canada.

Table 2.1 Recommendations of the Committee on National Statistics

General/primary researchers

1. Sharing data should be a regular practice.
2. Investigators should share their data by the time of publication of initial major results of analyses of the data except in compelling circumstances.
3. Data relevant to public policy should be shared as quickly and widely as possible.
4. Plans for data sharing should be an integral part of a research plan whenever data sharing is feasible.
5. Investigators should keep data available for a reasonable period after publication of results from analyses of data.

Secondary users

6. Subsequent analysts who request data from others should bear the associated incremental costs.
7. Subsequent analysts should endeavour to keep the burdens of data sharing on initial investigators to a minimum and explicitly acknowledge the contribution of the initial investigators.

Funders of social research

8. Funding organizations should encourage data sharing by careful consideration and review of plans to do so in applications for research funds.
9. Organizations funding large-scale, general-purpose data sets should be alert to the need for data archives and consider encouraging such archives where a significant need is not now being met.

Journal editors

10. Journal editors should require authors to provide access to data during the peer review process.
11. Journals should give more emphasis to reports of secondary analyses and to replications.
12. Journals should require full credit and appropriate citations to original data collections in reports based on secondary analyses.
13. Journals should strongly encourage authors to make detailed data accessible to other researchers.

Other institutions

14. Opportunities to provide training on data sharing principles and practices should be pursued and expanded.
15. A comprehensive reference service for computer-readable social science data should be developed.
16. Institutions and organizations through which scientists are rewarded should recognize the contributions of appropriate data-sharing practices.

Source: Adapted and reprinted from Fienberg et al. (1985: 22–32) with permission of the National Academy of Sciences, courtesy of the National Academies Press, Washington, D.C.

Support for the principle of data sharing and data retention has also been expressed by some professional organizations in their codes of research practice. For example, the American Sociological Association (ASA) has endorsed data sharing while recognizing that this practice may be limited by confidentiality agreements.[12] Thus its guidelines state:

(a) Sociologists share data and pertinent documentation as a regular practice. Sociologists make their data available after completion of the project or its major publications, except where proprietary agreements with employers, contractors, or clients preclude such accessibility or when it is impossible to share data and protect the confidentiality of the data or the anonymity of research participants (e.g. raw field notes or detailed information from ethnographic interviews).

(b) Sociologists anticipate data sharing as an integral part of a research plan whenever data sharing is feasible.

(c) Sociologists share data in a form that is consonant with research participants' interests and protect the confidentiality of the information they have been given. They maintain the confidentiality of the data, whether legally required or not; remove personal identifiers before data are shared; and if necessary use other disclosure avoidance techniques.

(d) Sociologists who do not otherwise place data in public archives keep data available and retain documentation relating to the research for a reasonable period of time after publication or dissemination of the results.

(e) Sociologists may ask persons who request their data for further analysis to bear the associated incremental costs, if necessary.

(f) Sociologists who use data from others for further analyses explicitly acknowledge the contribution of the initial researchers. (American Sociological Association, 1997: Section 13.05)

While these guidelines do not specifically state whether they apply to both quantitative and qualitative data, the implicit emphasis is on the former. However, the guidelines of some organizations, such as the Association of Social Anthropologists of the UK and the Commonwealth (1999) and the British Sociological Association (undated, c.2002) now include statements specifically relating to the sharing of the latter type of data (these guidelines are examined in Chapter 5).

Finally, the development of quantitative and, more recently, qualitative secondary analysis has been facilitated by the production of various guidelines for data donors on how to collect and prepare data for deposit in archives (see Inter-university Consortium for Political and Social Research, 2002; Marz and Dunn, 2000; Qualidata, undated a, undated b, undated c; Rolph, 2000). However, while there are a number of textbooks for quantitative researchers outlining resources and methods for doing secondary analysis of statistical data from social surveys (see Dale et al., 1988; Hakim, 1982; Kiecolt and Nathan, 1985; Riedal, 2000), there is relatively little equivalent information for qualitative researchers. In the United Kingdom, information on resources is now available through Qualidata's on-line catalogue of archived data sets (Qualicat)[13] but there is still very little methodological advice on how to go about re-using these and other sources of qualitative data.[14] In addition, as I show below in the discussions on the potential benefits and problems of re-using research data, the re-use of qualitative data has tended to be viewed as more problematic than that of quantitative data.

Potential benefits and problems of re-using data

The initiatives described in the previous section have been accompanied by an ongoing academic debate over the potential of data sharing and its corollary, secondary analysis, in the social, natural and bio-medical sciences. This debate began in the United States in the 1960s and was at its most intense in the 1980s when it focused on the possibilities and problems of opening up access to quantitative data (see Abel and Sherman, 1991; Baron, 1988; Ceci and Walker, 1983; Cecil and Boruch, 1988; Cherlin, 1991; Clubb et al.,1985; Colby, 1982; Colby and Phelps, 1990; Dale et al., 1988; Fienberg et al., 1985; Hakim, 1982; Hauser, 1987; Hedrick,1988; Hilgartner and Brandt-Rauf, 1994; Hogue, 1991; Hyman, 1972; Melton, 1988a, 1988b; Sieber, 1988, 1991; Stanley and Stanley, 1988; Sterling, 1988). Since the mid-1990s, there has been a growing focus in the United Kingdom and North America on the issues relating to the archiving and formal sharing of qualitative data following the establishment of Qualidata, and on informal data sharing and personal secondary analysis following pioneering work by qualitative researchers using these sources of data (see Alderson, 1998; Corti, 2000; Fink, 2000; Hammersley, 1997a; Heaton, 1998; Hinds et al., 1997; Hood-Williams and Harrison, 1998; Mauthner et al., 1998; Szabo and Strang, 1997; Thorne, 1994, 1998).

As Dunn and Austin (1998: 16) have pointed out, data sharing is a 'disputed norm' in scientific affairs. On the one hand, there is a view that primary researchers are best placed to analyse and safeguard data and that, as they collected the data, they should have first rights over publication. On the other hand, there is a view that the data generated by studies should in many cases be considered a public resource, especially if the work was funded by the government and/or if the information is in the public interest, and that such considerations should outweigh researchers' private claims to ownership and possibly stewardship of the data, as well as any commercial interest in withholding the data. Intersecting with these arguments are associated debates regarding research participants' rights to privacy and the associated agreements between researchers and scientists over the use of the data. Dunn and Austin (1998: 22) note that studies in which 'qualitative narrative interviews' are used pose particular problems with respect to confidentiality. More generally, the re-use of qualitative data is perceived to be more problematic than that of quantitative data, although not necessarily prohibitive. The potential benefits and problems associated with the re-use of both types of data are summarized below.

Data sharing is generally perceived to have many possible benefits in the social sciences. Those associated with the sharing of quantitative data include the potential for:

- reinforcing open scientific inquiry;
- facilitating the verification, refutation or refinement of the original results;
- promoting replication studies;
- facilitating the elaboration of earlier findings;
- facilitating the investigation of new research questions;
- developing theoretical knowledge;
- improving methods and analytical techniques;
- encouraging multiple perspectives;
- protecting against the use of faulty data;
- promoting comparative studies;
- promoting longitudinal studies;
- reducing research costs (compared to primary research);
- avoiding unnecessary duplication of primary research studies;
- promoting the more appropriate use of empirical data in policy-making;
- increasing opportunities for researchers to do empirical work;
- improving the standard of primary research;
- promoting cross-disciplinary and cross-boundary research;
- improving the generalizability of research using statistical methods;
- promoting analyses based on mixed methods.

At the same time, various problems and obstacles to sharing quantitative data sets have been identified,[15] including the:

- poor condition and documentation of data sets;
- efforts and costs of data sharing;
- loss of control over data;
- risk of criticism of the original work by peers;
- lack of credentials of some secondary users;

- incompatibility of computer hardware and software of data donors and data users;
- difficulties of accessing data;
- primary researchers' proprietary interests in the data;
- lack of incentives to donate data;[16]
- existence of disincentives to donate data;[17]
- risk to confidentiality;
- difficulties of obtaining informed consent.

Most of these claims apply across the social, natural and biomedical sciences, although there is one exception. That is, in the natural and biomedical sciences, researchers' proprietary interests in the data are more likely to be complicated by commercial concerns, as in the example of the Human Genome Project. This is less likely in the social sciences, where primary researchers are mainly concerned to retain ownership of the data for professional reasons – in order to be the first to develop analyses and publications based on the work, and for ethical reasons – in order to safeguard the data and protect the confidentiality of research participants (of which more below). Not only may these proprietary interests act against data sharing, at the same time, it has been observed that there are no real incentives for social researchers to share their data with others (see Baron, 1988; Clubb et al., 1985; Stanley and Stanley, 1988). In addition, social researchers have to consider the possibly conflicting interests of a wide range of stakeholders in the research, including the research participants, the research sites, the researcher's employers, the funders of the research, commercial groups, ethical committees and review boards, study advisory groups, archivists and archives, publishers, policy-makers, and professional associations.

Many of the above benefits of data sharing have been reiterated in the emerging literature on the secondary analysis of archived and other sources of qualitative data. Additional benefits have also been identified in relation to this type of data (see Fielding and Fielding, 2000; Hammersley, 1997a; Sandelowski, 1997). These include:

- avoiding over-burdening informants;
- facilitating more research on hard-to-reach groups;
- enabling additional research on sensitive topics;
- promoting the generalizability of qualitative research findings;
- salvaging data from primary work which was not completed.

However, in this and the wider literature on data sharing and secondary analysis, the re-use of qualitative data is generally perceived to be more problematic than that of quantitative data. While some archivists and qualitative researchers believe that the problems of sharing and re-using this type of data can be overcome, others are more sceptical or firmly opposed to this practice.[18] Three areas of particular concern are outlined below.

First, doubts have been expressed over whether secondary analysis of qualitative data is as cost-effective as that of quantitative data (Mauthner et al., 1998). Certainly the archiving of qualitative data can be particularly costly in terms of time and money; more physical space is also required in which to store the raw data, and the conditions have to allow for the preservation of degradable materials such as photographs and audio and video-tapes. While the costs of preparing data for archiving tend to fall upon the primary investigators and/or data archive, the users of this data have to meet the costs of finding appropriate data sets, checking the suitability of the data and meta-documentation, obtaining access to the data held by archives or third parties, re-producing and re-coding the data for analytical purposes, and re-analysing the data. There are also allied 'fieldwork' costs of visiting archives and, possibly, contacting members of the primary research team and/or re-contacting and obtaining informed consent from the original research participants. However, in contrast to this formal mode of data sharing, the informal sharing of qualitative data with colleagues is likely to be a cheaper and less time-consuming option; likewise the re-use of one's own qualitative data is likely to be less costly still.

Second, attention has been drawn to the difficulties of sharing data while respecting confidentiality agreements with research participants. This is an issue in quantitative as well as qualitative research. For example, while quantitative data can be relatively easily anonymized or aggregated, in theory linkages could be made between data sets, or data could be dis-aggregated, thereby allowing individuals to be identified. Concern has also been expressed about the security of data sets containing sensitive information.[19] However, the issue of confidentiality is seen as posing particular problems for the sharing of qualitative data (Corti et al., 2000; Dunn and Austin, 1998). Qualitative data can be anonymized but this is an expensive and time-consuming process and is not guaranteed to prevent informants from being identified from contextual information. Efforts to disguise the identities of informants may also lead to distortion of the data, rendering meaningful secondary interpretation impossible. Accordingly, other strategies have been developed or proposed for dealing with this issue. For example, the Institute for History and Biography in Germany does not anonymize data but requires users to sign confidentiality agreements preventing them from disclosing the identity of individuals or organizations (Leh, 2000).[20]

Finally, various epistemological concerns have been raised about the re-use of qualitative data. For example, while data sharing and data retention have been promoted as ways of enabling empirical work to be independently examined, it is a matter of debate as to whether qualitative studies are amenable to verification through re-analysis in the same ways as studies based on quantitative methods (Corti, 1999; Hammersley, 1997a). As Hammersley (1997a: 132) has observed, in practice the closest qualitative researchers have traditionally come to verifying studies is through conducting additional primary research designed to emulate the original, such as Lewis' (1951) re-study of Redfield's (1930) anthropological work on a Mexican village. However, some qualitative researchers have expressed support for the principle of verification, such as Scheff (1986) who has argued that there is a need to keep 'living texts' (such as video and sound recordings) in order for other researchers to assess the validity of analyses and Harvey Sacks who based conversation analysis on this principle (see Chapter 4).

Concerns have also been raised about whether it is tenable for qualitative data to be re-used in new studies. One issue is the degree of 'fit' between the data and the secondary study question (Hinds et al., 1997; McArt and McDougal, 1985). Whereas statistical data, particularly of the type generated through omnibus social surveys, is regarded as a resource for multiple research studies, qualitative research data sets are seen as a 'rich' but nonetheless dedicated resource, collected for a particular set of primary research aims and objectives. Whether these data 'fit' with other secondary research questions, particularly in terms of the depth and breadth in which qualitative studies explore topics, is open to question. Moreover, because secondary research relies on data which were not purposively collected for the study, qualitative secondary studies could be perceived to be inferior to primary research. This view contrasts with research based on the use of existing quantitative data, where secondary analysis is regarded as the methodology of choice for examining questions which require high quality, large or longitudinal, statistical data sets which it would be relatively difficult, expensive and impractical to investigate using primary methods in all cases.

Another key issue is whether qualitative data can be examined outside of the context in which it was collected. In qualitative research, the interpretation of data is generally perceived to be dependent on the primary researcher's direct knowledge of the context of data collection and analysis obtained through their own personal involvement in the research (Corti, 2000: para 26; Heaton, 1998). From this perspective, data collection and analysis are best conducted as contiguous processes, at the height of the researcher's involvement with the research. While some archivists and researchers believe that meta-documentation of the research process can compensate for not 'being there', resistance to this view has been identified as one of the obstacles to qualitative data sharing in the United Kingdom (Corti, 1999, 2000).

The perceived problems with the sharing of qualitative data, together with the limited availability of archived qualitative data sets, may well explain the greater emphasis placed by North American researchers in particular on the potential of qualitative secondary analysis involving the use of researchers' own data sets (either solely, or in informal collaboration with colleagues who may or may not add their own data to the pot). In this situation, the primary-cum-secondary researcher is seemingly in a good position, because of his or her familiarity with the data, to define a secondary study question which is apposite.[21] However, doubts have been expressed about whether or not qualitative researchers can meaningfully re-use their own data as well as other researchers' data (Mauthner et al., 1998). These and other issues will be examined in more depth in the remainder of the book.

Conclusion

In this chapter I have suggested that the emergence of secondary analysis as a qualitative methodology has been shaped by wider developments in data archiving and data sharing and by a growing debate by social researchers as to the potential opportunities and problems in sharing and re-using quantitative and qualitative data sets. On the one hand, the development of qualitative secondary analysis has been helped by archiving strategies which have been extended to include increasing numbers of qualitative data sets, and by the introduction of various official policies and guidelines which have begun to encourage researchers to use pre-existing qualitative as well as quantitative data where this is feasible, as well as to archive qualitative data arising from their primary research work. On the other hand, the development of qualitative secondary analysis has been limited by the general perception that the secondary analysis of qualitative data is more problematic than that of quantitative data. While some archivists and researchers believe that these problems can be overcome, some researchers do not feel that qualitative data can be re-used in the same ways as quantitative data for a variety of technical, ethical and epistemological reasons.

These conflicting views on the potential benefits and problems of qualitative secondary analysis raise interesting empirical questions about the extent to which researchers have attempted to make use of secondary qualitative data, how they have so far approached 'doing' qualitative secondary analysis, and whether or not they found it useful. In the next chapter, I review a number of qualitative secondary studies and examine how the methodology has been used in practice. The issues arising from this review will then be discussed in more detail in the remainder of the book, including consideration of the implications of the development of qualitative secondary analysis for our understanding of the nature of qualitative research.

Notes

1. Further information on the HRAF project is available at: http://www.yale.edu/hraf/ [accessed 11/3/2003].
2. Other qualitative researchers who deplore the HRAF approach include Vidich and Lyman (1994: 28) who have argued that 'The trait data in the Yale cross-cultural files represent ethnography in a form disembodied from that of a lived social world in which actors still exist. They are a voluminous collection of disparate cultural items that represent the antithesis of the ethnographic method'.
3. Further information on the Murray Research Center is available at: http://www.radcliffe.edu/murray/ [accessed 11/3/2003].
4. Information on data archives around the world is available at the Council of European Social Science Data Archives (CESSDA) website at: http://www.nsd.uib.no/cessda/ [accessed 11/3/2003].
5. Further information on the Mass-Observation archive is available at the University of Sussex's library website: http://www.sussex.ac.uk/library/massobs [accessed 11/3/2003].
6. At the time of writing the ESRC was in the process of reviewing its policy and arrangements for the archiving of quantitative and qualitative data. In 2003 Qualidata became part of the Economic and Social Data Service (ESDS) at the University of Essex. Qualidata's web pages are in the process of being transferred to http://www.esds.ac.uk/qualidata [accessed 31/12/2003]. (Boddy, 2001).
7. For a list of archives which Qualidata has used, see Corti (1999: 19).
8. Further information on Qualidata and Qualicat is available at: http://www.qualidata.essex.ac.uk/about/introduction.asp [accessed 11/3/2003].
9. It was even suggested that here data sharing should occur before results were published in order to allow for independent verification of the findings first (Fienberg et al., 1985: 11).
10. This method of sharing data was also advocated by Hauser who, in a guest editorial of the *American Sociological Review*, called for journals to improve scientific communication by producing machine-readable supplements, including:

> Technical appendices, documentation of data, and supporting data, either in raw form (in the case of small bodies of data or survey extracts) or in the form of cross-classification tables and/or moments matrices. (Hauser, 1987: vii)

11. For example, in the United Kingdom, the Medical Research Council's (2000) guidelines on good research practice recommend that raw data is kept for ten years, as do The Wellcome Trust (2002).

12. Similarly, the Social Research Association (SRA) recognizes that the re-use of pre-existing data can reduce the burden on subjects, but it also highlights the issues of informed consent and confidentiality and how they may limit this practice:

> One way of avoiding inconvenience to potential subjects is to make more use of available data instead of embarking on a new inquiry. For instance, the preferred option would be to make greater statistical use of administrative records by conducting secondary analysis of existing data for which informed consent had been granted. By linking existing records, valuable social research information may be produced that would otherwise have to be collected afresh. But there are often issues of confidentiality in linking records which may affect what can be done. Individual subjects should not be affected by such uses provided that their identities are protected and that the purpose is statistical, not administrative. On the other hand, subjects who have provided data for one purpose may object to its subsequent use for another purpose without their knowledge (see clauses 4.3 iii, 4.6 and 4.7). This is particularly sensitive in the case of identified data. Decisions in such cases have to be based on a variety of competing interests and in the knowledge that there is no "correct" solution (see clause 4.4). (Social Research Association, 2002: Section 4.1)

13. Qualicat can be accessed on the Internet at: http://www.qualidata.essex.ac.uk/about/introduction.asp [accessed 11/03/2003].

14. Even textbooks on life histories and oral history tend to focus on the collection and use of these data in the context of primary rather than secondary research (see Atkinson, 1998; Howarth, 1998; Langness, 1965; Langness and Frank, 1981; Miller, 2000; Thompson, 2000b). Thus they mainly advise on how to collect, record and, in some cases, prepare data arising from primary research for archiving, rather than how go about doing research using secondary resources. Exceptions are Konopásek and Kusá (2000) who discuss the re-use of life stories in an ethnomethodological study, and Hill (1993) who explores the value of archives for doing 'sociobiographical and sociohistorical' analyses.

15. Some of the technical issues referred to were raised in the 1980s and have subsequently been reduced as a result of advances in computing.

16. One possible incentive, proposed by Baron (1988), is that primary researchers who donate their data should be credited for an 'assist' with any subsequent work.

17. These include the possibilities of errors being identified, the material being mis-used, and the perceived poorer status of research based on secondary rather than primary research methods (Estabrooks and Romyn, 1995; Fienberg et al., 1985).

18. In a survey of researchers' attitudes towards re-using archived data in 2000, Qualidata found that only 63 out of a total of 542 academics stated categorically that they would not consider using archived qualitative data sets; however, only one-third of respondents had attempted to use secondary resources (Corti, 2000; Thompson, 2000a). Similarly, in interviews with 70 anthropologists, Jackson (1990) also found this group to have ambivalent attitudes towards sharing their fieldnotes with others.

19. See, for example, the debate surrounding the establishment and use of the Icelandic Health Sector Database (HSD) created by deCode (Rose, 2001).

20. James and Sørensen (2000) also report that the Murray Research Center in the United States was rethinking its policy on de-identification and considering alternative strategies, such as the careful screening of users.

21. It has also been argued that it is incumbent upon primary researchers to make full use of *all* the data supplied by research participants (in quantitative and qualitative research) even if it is not immediately relevant to the terms of the primary research (Clayton et al., 1999).

3
Types of Qualitative Secondary Analysis

Overview of the studies	36
Supra analysis	39
Supplementary analysis	41
Re-analysis	45
Amplified analysis	47
Assorted analysis	50
Conclusion	51
Notes	52

As we have seen from the two previous chapters, qualitative secondary analysis is a new and emerging methodology. In this chapter I examine how social researchers have begun to use the methodology in practice. What types of qualitative data have they drawn on? Whose data sets were used? Were the studies undertaken to explore new research questions or to verify previous work? To what extent have the potential benefits and problems of re-using qualitative data been realized in practice? The chapter builds on a review of qualitative secondary studies carried out in the area of health and social care research (Heaton, 2000).[1] Full details of the background and conduct of the review are described in Appendix A. Here I describe the characteristics of the studies and distinguish different types of qualitative secondary analysis that have been employed in this area of social research. Epistemological, ethical and methodological issues arising from these findings are discussed in more depth in subsequent chapters.

Overview of the studies

Characteristics of the studies

Sixty-five studies were identified which involved the use of pre-existing qualitative research data either to investigate new research issues or to re-examine primary study questions for purposes of verification.[2] Of these studies, 41 were clearly defined by the authors as being based on secondary analysis. The methodology used in the remainder was either undefined or described using other terms such as *'post hoc'* analysis, 'reanalysis' and 'retrospective latent content analysis'. Fifty-nine of the studies were published post-1990 and 19 were published before 1994, pre-dating the first methodological papers focusing exclusively on the secondary analysis of qualitative data. The majority of the studies (51 or 78 per cent of those reviewed) were carried out by authors working in North America. Of the remaining studies, 12 were by authors based in the United Kingdom, one originated from Sweden, and one was a pan-Canadian and Swedish study. While allowing that these findings to some extent reflect the geographical coverage of some of the resources searched and the limitations of the search strategies adopted (which excluded studies not published in English), the results suggest that there is particular interest in secondary analysis as a qualitative methodology in the United States and Canada, from where much of the methodological work on the topic has also been derived (see Chapter 2).

Of the 65 studies, 36 were carried out by researchers who had re-used their own previously collected data sets (see Table 3.1). Another 20 studies were published by one or more authors who had been involved in the primary research, in collaboration with authors who had not. In one of these studies, a mix of auto- and independent data sets were re-used. In only nine of the studies did the researchers have no apparent involvement with the original research. These results are important because they indicate that the majority of studies in this area (86 per cent) have been undertaken by researchers with some first-hand knowledge of the context in which the data were originally collected and analysed, while the remainder were by authors who had no direct knowledge or experience of the original study from which the data were derived.

Single data sets were re-used in 48 studies (74 per cent). In some cases, authors had re-used the same data set on more than one occasion, resulting in two or more secondary studies. Fourteen studies had drawn on two data sets, and three utilized three or more data sets in total. The data sets varied in size, for example, they ranged from under ten to over 200 interviews. In around half the studies the primary data set(s) were re-used in

Table 3.1 Authors' relationship to the primary data set(s)

Relationship	Number (percentage)
Previously involved with the primary research	36 (55%)
Mixed (some authors were involved with the primary research and some were not; or a mix of independent and auto-data sets were used)	20 (32%)
No previous involvement with the primary research	9 (14%)

full and in the other half part of the data set was selected for secondary analysis. The studies involved the re-use of various types of research (non-naturalistic) data and nine studies also involved the collection and analysis of additional primary data or pre-existing naturalistic data, such as autobiographies.[3] Face-to-face interview material was the main type of qualitative data utilized in 52 studies; other forms of qualitative data were also drawn upon, including those derived from observational work, focus groups, surveys, telephone interviews, vignettes, comments on questionnaires, fieldnotes, taped social interaction, and published ethnographies.

Types of qualitative secondary analysis

Previous efforts to distinguish different types of qualitative secondary analysis have been mainly theoretical (see Thorne, 1994, 1998) or based on the review of individual studies (see Hinds et al., 1997). The present, more recent and extensive review of studies published in the health and social care literature provides an opportunity to develop a typology based on how the methodology has been used in practice.

In one respect the studies were found to be surprisingly uniform: all were carried out in order to investigate new or additional research issues, except for one study which aimed to verify the original research findings.[4] However, the studies varied in terms of how far they diverged from the primary research. Some focused on new questions and employed theoretical perspectives which clearly differentiated them from the primary research. Others explored additional questions or issues in ways which added depth to and embellished the original research. The studies also varied in terms of the range of data drawn upon. In some secondary studies just one data set was re-used, while others utilized data arising from two or more primary studies, and yet others used a mix of pre-existing qualitative data sets, naturalistic qualitative material and/or data from primary research conducted in conjunction with the secondary work.

Table 3.2 Types of secondary analysis of qualitative data

Supra analysis	Transcends the focus of the primary study from which the data were derived, examining new empirical, theoretical or methodological questions.
Supplementary analysis	A more in-depth investigation of an emergent issue or aspect of the data which was not considered or fully addressed in the primary study.
Re-analysis	Data are re-analysed to verify and corroborate primary analyses of qualitative data sets.
Amplified analysis	Combines data from two or more primary studies for purposes of comparison or in order to enlarge a sample.
Assorted analysis	Combines secondary analysis of research data with primary research and/or analysis of naturalistic qualitative data.

Based on this information, five varieties of secondary analysis were discerned (see Table 3.2). The first two types distinguish those studies where secondary analysis was used to investigate questions which transcend the primary research (supra analysis) from those where the methodology was used to extend the primary research (supplementary analysis). The third type (re-analysis) differs from these varieties of secondary analysis in that it was concerned not with investigating new or additional questions but with verifying the results of prior studies. The two remaining types further distinguish studies in which secondary analysis was conducted using multiple qualitative data sets (amplified analysis) or using qualitative material derived from a mix of sources (assorted analysis).[5]

This typology refines and extends previous models of classification, and hence employs new terms to describe the different varieties of secondary analysis identified. As it was mainly derived from a review of studies from one field of research, it is not necessarily representative of the use of the methodology in other areas of social research. It is also important to note that these are ideal types and hence individual studies may not always exhibit all the characteristics of a given type. In addition, the types are not exclusive: that is, supra, supplementary and re-analysis may all be conducted using single, multiple or mixed data sets and hence may share the characteristics of amplified and assorted analysis. In the following, more detailed, description of each of these varieties of secondary analysis, the examples are classified according to their principal characteristics.

Supra analysis

In 'supra analysis' the terms of the primary study from which the data were derived are transcended. Such studies involve the investigation of new theoretical, empirical or methodological questions.[6] This type of qualitative secondary analysis shares some characteristics with a variety of secondary analysis called 'armchair induction' identified by Thorne (1994, 1998). However, supra analysis differs in that it may be conducted by the same researchers who carried out the primary research and is not necessarily restricted to theoreticians. Fourteen of the 65 secondary studies included in the review developed analyses which went beyond the terms of the primary work, focusing on other aspects of the data and often employing a new theoretical perspective. An example of such a study is provided by Weaver (1994). She drew on Roth's then 40 year-old fieldnotes from his observations while an attendant in a tuberculosis sanatorium. In what is described as 'a well-nigh unique example of secondary ethnographic data analysis' (1994: 77) she examines the disease from the perspective of cultural theory.

I also undertook a secondary study in which I examined pre-existing data from a new perspective (Heaton, 2001). The aim of the primary study, with which I was involved, was to examine the effectiveness of hospital discharge procedures for younger physically disabled adults from the perspective of informal carers (Heaton et al., 1999). In qualitative interviews, carers discussed their experiences of the discharge process and the adequacy of continuing care arrangements at home. Subsequently, in what began as a separate doctoral research interest, I examined the temporal organization of hospital discharge procedures from the Younger Disabled Units and other specialist care units in the original study. The secondary study was set in the context of the previous literature on the temporal organization of care in hospitals and in community settings, and the analysis considered the extent to which the discharge procedures bridged the temporal gap between the rhythms and routines of hospital and home-based care regimes. In these respects, the secondary study diverged from the primary research, employing a temporal perspective through which new aspects of the data were explored. Similarly, another of the secondary studies reviewed also explored the 'temporal landscape' of night nursing using data the authors had formerly collected to examine night nurses' attitudes to working practices (Brown and Brooks, 2002).

Although not defined as an example of secondary analysis by the authors, a study by Bloor and McIntosh (1990) arguably exhibits the characteristics of both supra and amplified analysis. They developed what they conceptualized as a *'post hoc'* analysis of data from two separate studies of health visiting and therapeutic communities, the focus of which was not planned at the outset of the study design and data collection:

> Of course, the health visiting and therapeutic communities studies were not undertaken in order to compare professional-client relationships in different types of service-provision, nor were they undertaken to elaborate a Foucauldian approach to client resistance. The possibility of writing this chapter only occurred to us *post hoc*, when we realized that we both had data bearing upon issues of power and contest which showed both similarities and dissimilarities in techniques of client resistance. Had we intended from the outset to produce a typology of client resistance we would have designed different studies, employing more readily comparable methods of data collection and possibly focusing on different research settings. Nevertheless, we feel that the analysis of client resistance produced here, albeit a mere by-product of *post hoc* comparisons of independent studies, may have some value, both in terms of its applicability to some other forms of professional-client relationship, and in terms of its uniqueness as a typology of client resistance. (Bloor and McIntosh, 1990: 161–62)

In focusing on forms of surveillance in professional-client relationships and associated strategies of resistance, Bloor and McIntosh develop a new, free-standing analysis which clearly goes beyond the terms of the original studies.

While the above studies applied new theoretical perspectives on culture, time and power to develop further analyses of pre-existing data, supra analysis has also been used to explore new empirical questions arising from primary research. A good example of this is provided by Clayton, et al. (1999). In their secondary study they examined the unsolicited comments provided by respondents with multiple sclerosis in a survey conducted as part of their primary quantitative study of health promotion and quality of life in chronic illness. Of the 811 people who participated in the study, a quarter added extensive comments on the 36 page questionnaire booklet. A secondary analysis of this qualitative data was undertaken in order to understand what the respondents had written in the margins of the questionnaire that was so important to them. Three new research questions were posed: were there any differences in the characteristics of respondents who added comments compared to those who did not? What categories and themes were apparent in the comments? And why did respondents go to the trouble of answering unasked questions despite having writing difficulties associated with multiple sclerosis? In order to examine these questions, the authors used Strauss and Corbin's (1998) micro-analytic method to define categories and re-code the data.

Similarly, in three secondary studies, the focus shifted to informants' use of metaphors in their accounts of medical encounters. Thus, Jairath

(1999) examined a subset of patients who had used metaphorical language to describe their pain in a previous clinical study of patients' responses to stair climbing two weeks post myocardial infarction. Jenny and Logan (1996) examined the meaning of metaphors used by critical care patients in describing their ventilator weaning experience. And, in a rare example of a secondary study in which the published findings of a qualitative study were re-used (and not the raw data itself), Pascalev (1996) explores two metaphors of death and dying used by medical professionals in Intensive Care Units as reported in Zussman's (1992) text.

In addition to the various types of qualitative data used in the above studies, two secondary studies used pre-existing oral histories as the basis for their secondary studies. Bevan (2000) used archived data from a project entitled 'The oral history of General Practice in Britain, 1935–52', held at the British Library National Sound Archive (and elsewhere), to examine the influence of the family in general practitioners' decisions to follow this career. Bloor (2000) used oral history interviews from the 1970s, deposited in the South Wales Miners' Library, to explore lay understandings of, and collective responses to, pneumoconiosis (Miners' Lung) in the 1920s and 1930s. Other examples of substantive supra analysis from the health and social care literature include: a study examining the relationship between primary care physicians' philosophies and styles of practice and the 'history of ideas' in preventive medicine in which the authors re-used data from their primary study on the delivery of preventive services (Aita and Crabtree, 2000); and a study of attrition in two ethnographic longitudinal field studies of drug users in which qualitative data were re-examined using quantitative methods (Fendrich et al., 1996).

Finally, this variety of secondary analysis also includes a small group of studies in which data were re-used to explore methodological issues. These demonstration projects include work by Weaver and Atkinson (1994) where they re-used Roth's data for the purposes of illustrating a book on micro-computing and qualitative data analysis. In another study, Atkinson (1992) re-used data from his doctoral research in order to explore different ways of reading and re-reading data, comparing the orthodox 'cut and paste' method of analysis with that of a narrative-based approach. Similarly, the value of biographical analysis using 'thematic field analysis' and 'microanalysis' was also explored in a study by Jones and Rupp (2000).

Supplementary analysis

'Supplementary analysis' involves a more in-depth focus on an emergent issue or aspect of the data which was not addressed, or was only partially addressed, by the primary research. Unlike supra analysis, it is more closely related to the analytical remit of the primary study, extending, rather than exceeding, the original work. For example, the focus of the secondary analysis may shift to a particular theme, or to issues pertaining to a sub-sample of the primary study population. Supplementary analysis is in these respects related to Thorne's (1994, 1998) 'retrospective interpretation' and 'analytic expansion' categories (the former referring specifically to researchers' re-use of their own data) and other examples of this kind of secondary analysis have been described previously (Heaton, 1998; Hinds et al., 1997). As the foci of supplementary analyses are compatible with that of the primary work, the two may be difficult to separate, particularly where researchers report selectively the results of primary research in multiple publications rather than in a single, overarching, publication. However, selective reporting differs in that it occurs in the context of the ongoing primary research. By contrast, supplementary analysis examines in more depth a theme or sub set of the data which has emerged as a *post hoc* matter of interest.

Supplementary analysis was found to be by far the most common form of qualitative secondary analysis in the health and social care literature, exemplified by 39 out of the 65 studies identified (60 per cent). Around half these studies re-used a part of the material from the primary data set(s). The basis for these sub-sets varied and was not always clearly stated. In some cases, sub-sets were comprised of data relating to a particular group from the original study sample. For example, McLaughlin and Ritchie's (1994) secondary study of the experiences and circumstances of ex-carers involved the re-use of both quantitative and qualitative data from a larger primary study of Invalid Care Allowance (ICA) claimants. Likewise, in one of their two secondary analyses of data from their primary study of women cocaine users, Kearney et al. (1994a) focused in more depth on the mothers in this sample. In other cases, sub-sets were selected according to their relevance to the particular thematic or substantive focus of the secondary study. For example, Powers (1996) focused on women's relationships in a nursing home, based on a wider study of social networks in this context; Clarke-Steffen (1998) re-used data relating to one of the children's cancer settings featured in a primary study of pediatric oncology nurses' peak and nadir experiences; and in their secondary study of parenting in families where one parent has a chronic illness, Rehm and Cantanzaro's (1998) focused on a case study sample of 23 families who were followed up each year for four years as part of a larger longitudinal study of 604 families. In other studies, sub-sets of the data were selected from a particular phase of the data collection or point in a series of interviews with informants (Knafl et al.,1995; Szabo and Strang,

1999) and for their relevance to the particular focus of the secondary analysis (Deatrick et al., 1992; Gallo and Knafl, 1998).[7]

In the other half of these studies, the entire data set was re-used. These secondary studies vary in how easily they may be distinguished from the related primary research. At one end of the continuum the distinction is relatively plain and the secondary studies are recognizably free-standing. For example, a secondary study by Vallerand and Ferrell (1995) focused on the concept of control, following on from one of the author's exploratory, descriptive study of pain management at home by cancer patients. For the secondary study, each of the authors independently reviewed the transcripts and coded responses specifically to examine this issue. Similarly, in their supplementary analysis of data from a study on the process of weaning patients from mechanical ventilation, Logan and Jenny (1990) also individually analysed the data, agreed a new coding framework and shared their analysis of a new nursing diagnosis – 'dysfunctional ventilatory weaning response' – with the study informants. At the same end of the continuum, Princeton (1993) undertook a secondary analysis of data from her 'exploratory and descriptive study' of the educational preparation, workload and anticipated career patterns of nurse administrators in higher degree nursing programmes. The secondary study focused on the 'knowledge domains' specific to the nurse administrator role and the implications for their role preparation. The distinctive focus of this study is indicated by the author's comment that:

> In reviewing the interview data collected during the earlier study, two important yet different research questions related to role preparation became apparent that had not been posed by the researchers, but which the participants addressed voluntarily during telephone interviews... (Princeton, 1993: 61)

Similarly, following on from their research on employment, maternal and spousal role experiences and the relationship to health outcomes, the authors observed that:

> it became apparent that the construct of role integration needed further development. The logical next step was to extend this prior work with an in-depth analysis of the qualitative interview data from the study to gain a clearer understanding of the aspects, processes, and patterns of role integration. (Hall et al., 1992: 448)

These examples illustrate how, in supplementary analysis, the focus is on issues which the original research was not designed to address directly but which nevertheless emerge as relevant concerns in the course of the work.

A further example of this type of secondary analysis is provided by a secondary study where data was re-used from the authors' previous research on the impact of nursing intervention on self-care and on the morbidity of patients receiving chemotherapy (Messias et al., 1997). In the secondary study, which is described as an example of Thorne's 'analytic expansion' variety of secondary analysis, the authors examined a topic – patients' perspectives on fatigue while undergoing chemotherapy – which was not specifically addressed by the interview guides but which was raised by the informants. Katz (1997) also developed a supplementary analysis of the experience of disclosing HIV infection to family and friends from her previous study of the consequences of uncertainty in the lives of people with HIV infection. As she notes:

> While the experience of disclosing HIV seropositivity to family and friends was not the primary purpose of the study, all the participants spoke of their personal reaction to the news of their serostatus and the response of family and friends. (Katz, 1997: 138)

Other examples of the spawning of this type of secondary analysis are provided by Sandelowski and Black (1994) who drew on their study of the transition to parenthood of infertile couples to develop a secondary study examining parents' perceptions of the relationship between them and the fetus. Sandelowski also re-used the same data set to develop a supplementary analysis of the influence of fetal ultrasonography on expectant mothers' and fathers' experience of pregnancy (Sandelowski, 1994a) and a supplementary analysis of couples' experiences of fetal ultrasonography as a type of 'technology transfer' (Sandelowski, 1994b).

Towards the centre of the continuum, where the primary research stops and the secondary research starts becomes harder to discern. For example, in one of two secondary analyses of the same data set, Kearney, et al. (1994b) developed a *'post hoc'* analysis using the entire data set. Whereas the purpose of the primary study was to describe the influences on and effects of cocaine ('crack') addiction in women, the supplementary analysis examined women crack smokers' views on safer sex and contraception. As the authors explain:

> Although fertility and birth control were not among the primary research questions during data collection, the life-story format used in all interviews brought out from all the participants tales of sexual relationships, pregnancies, and contraception experiences – topics that were central to many of the women's life experiences. As the study progressed, content related to sexuality and childbearing was granted more attention because of its centrality in many participants' stories, but analysis of these topics did not begin until after the data set was complete. (Kearney et al., 1994b: 145)

Thus, some 'attention' was given to this topic during the primary data collection although analysis only commenced after all the data were collected.

Moving on to the other end of the continuum, some secondary studies were more difficult to distinguish from their parent studies because they re-used all the primary data and had similar aims to the primary research. For example, although Jones (1997) defines her study of women's representations of menopause as a 'secondary analysis', she notes that: 'The overall purpose of both the original study and the secondary analysis was to increase understanding of women's experience of menopause in order to improve the quality of health care to this population' (1997: 59). The main difference in the studies, she argues, is that whereas the primary study examined women's experiences of menopause, the secondary analysis examined their representations of that experience. In another example, Breckenridge (1997) re-used data from a study of the perceptions of clients with end-stage renal disease on decision making regarding dialysis treatment options. The secondary study appeared to have similar aims, focusing on the factors that influenced decisions regarding types of dialysis treatment, using content analysis for this purpose. Finally, in a secondary study of nurses' views on elderly clients' mobility by Rush and Ouellet (1997), the authors focus on one of the two questions the primary study was designed to address.

Re-analysis

This type of secondary analysis differs from the first two types in that pre-existing data are re-examined to see if they support the original interpretations rather than to address new research questions or issues. Through 're-analysis', previous research findings may be confirmed and validated, or they may be questioned and refuted.

While this type of secondary analysis is generally recognized at a theoretical level, only one example of such a study was found in the health and social care literature. Popkess-Vawter et al. (1998) conducted a self-defined secondary analysis of data on overeating among women who 'weight-cycle' which the first author had completed. As they explain, the aim of the primary study was to examine the basis of overeating episodes in this group using 'reversal theory' whereas the purpose of the secondary analysis was:

> to analyze further subjects' interview responses to ensure that no important information was omitted in the primary analysis, which used a reversal theory coding system only. The secondary analysis provided a validity check for the primary coding results and an accuracy check for complete interpretation. Using methodological triangulation ... two methods of viewing the same empirical content, two coders performed a content analysis of the interview data with no consideration for reversal theory. (Popkess-Vawter et al., 1998: 71)

In this secondary study, alternative methods of analysis were used in a form of methodological triangulation to check the original analysis. This may be compared to researchers using new statistical techniques for analysing quantitative data sets to verify findings obtained using other methods of analysis.

The lack of re-analyses in the health and social care literature provides some support for the view that few studies in social science are designed to corroborate previous work (see Chapter 1). A wider trawl of qualitative studies in other areas of social research revealed only a few examples of re-analyses conducted as part of longitudinal studies and re-studies. For example, Walkerdine and Lucey (1989) followed up Tizard and Hughes' (1984) study of the roles of mothers and teachers in children's language and cognitive development. They re-analysed transcripts from the original study and also conducted further longitudinal research on the sample of children involved. Working from a different perspective, Walkerdine and Lucey proposed a different interpretation of the data.

In community studies and anthropology, pre-existing research has been used in re-studies in two different ways. In one group of re-studies, the published findings provided the baseline for the follow-up research to assess social change. Here neither the findings nor the raw data are revisited in order to verify the original research. Examples of these kind of re-studies include: the Lynds' *Middletown in Transition*, a re-study of their previous *Middletown* (Lynd and Lynd, 1929, 1937), which was also followed up later by Caplow et al. (1982, 1983); Burgess' study (1981, 1983) and re-study (1987) of Bishop McGregor School; Earle et al.'s (1976) re-study of religion and social change in Gastonia, following on from Pope's (1942) study; Devine's (1990, 1992) qualitative re-study of the *Affluent Worker* series (Goldthorpe et al., 1968a, 1968b, 1969); Stacey's study (1960) and re-study of Banbury (Stacey et al., 1975); and Gallaher's (1964) *Plainville Fifteen Years Later* which succeeded West's (1945) *Plainville, USA*.

This contrasts with the second group of re-studies in which the original research is challenged by independent researchers, mainly on the basis of their critical evaluations of the published findings (rather than the raw data) together with their further primary research. Examples of these critical re-studies include: Freeman's (1983) *Margaret Mead and Samoa* in which he contests Mead's (1928) work and that of another ethnographer's re-study of her work in the 1950s (Holmes, 1957); Lewis's (1951) *Life in a Mexican Village: Tepoztlan Restudied* in which he re-studied and re-interpreted Redfield's (1930) original research; and Boelen's (1992) re-evaluation of William Foote Whyte's (1993/[1943]) *Street Corner Society*, based mainly on her follow-up interviews with informants from the same community between 1970 and 1989. Although these re-studies did not involve the re-analysis of the 'raw' data sets, the authors did re-analyse the published findings and also consulted with the primary researchers about the original work.

Moreover, in response to Boelen's critique published in the same special issue of the *Journal of Contemporary Ethnography* devoted to *Street Corner Society*, Whyte (1992) describes how he obtained and re-read his own fieldnotes and correspondence with the informants in the study, which he had deposited at the library of the New York State School of Industrial and Labor Relations. Whyte used evidence from this material in support of his original analysis (adding that the material would be returned to the archive and hence be available for others to check for themselves). Whyte had hoped that the other social scientists asked to comment on Boelen's text in the same issue of the journal would appraise their respective interpretations and support his findings. However, to his apparent dismay, Vidich (1992), Richardson (1992) and Denzin (1992) refused to be drawn into a debate over who was right and who wrong. Indeed, Denzin ends his article by asking the question: 'Do we want the kind of classic sociology that Whyte produced and Boelen, in her own negative way, endorses? (Denzin, 1992: 131). As Whyte (1997: 31) later commented, his view is that there are 'social and physical facts' and that Denzin's so-called 'deconstructivist critique' is a sociological 'dead-end street'. For Whyte, then, the form of 'repeat' analysis he conducted of his own data, and which could in principle be performed by independent analysts, allows researchers to double-check the veracity of qualitative research findings.

On the basis of this wider literature it would therefore appear that re-analyses of qualitative data sets and published studies are not common. Those which have been conducted also vary as to whether they draw only on published findings or revert to the raw data as well, and whether or not they involve additional primary research. However, given the small number of such studies, they are collapsed into one generic 're-analysis' category for the purposes of this typology.

Amplified analysis

In addition to distinguishing types of secondary analysis on the basis of the degree of divergence between the questions respectively addressed by the secondary and primary studies, secondary studies can be further distinguished according to whether or not they re-used one or more data sets. Thus, as we have seen, the supra analysis by Bloor and McIntosh (1990) involved the use of multiple data sets. These are examples of 'amplified analysis'.

This type of secondary analysis is generally conducted in order to examine common and/or divergent themes across data sets. Amplified analysis can involve the comparison of different study populations or the pooling of data on a similar population. This definition incorporates similar approaches previously described as 'amplified sampling' (Thorne, 1994, 1998), 'aggregated analysis' (Estabrooks et al.,1994), and 'pooled case comparison' (West and Oldfather, 1995), which all involve the secondary analysis of multiple pre-existing data sets. Amplified analysis was the second most common type of secondary analysis identified by the review of the health and social care literature. Of the 17 examples found, 14 made use of two data sets. All these studies were conducted by authors who had been involved with the primary research, although additional independent researchers also collaborated with some of the secondary studies.

Two main groups of amplified analyses were discerned. In the first, researchers who separately conducted independent studies later performed an analysis of the two data sets to further explore common and/or divergent issues across two study populations. For example, Yamashita and Forsyth learned, through meeting at a conference, that they had each conducted studies of families' reactions to a relative's mental illness while working, respectively, in Canada and the United States. This meeting led to a secondary study in which they examined similarities and differences in the two data sets (Yamashita and Forsyth, 1998). Using 'aggregated analysis' (Estabrooks et al.,1994), the authors produced a synthesis of the findings from the two studies. In addition, Backett and Davison (1995) published what they defined as a 'comparative analysis' of two data sets they had separately collected. Recognizing that both data sets examined people's perceptions of health and illness in the context of everyday life, the authors subsequently analysed the discourses of physical and social aging and the implications for health promotion using these data sets. Like Bloor and McIntosh's (1990) secondary study of therapeutic communities, this study also shares some characteristics of supra analysis.

Other examples from this group, with more of an emphasis on comparing differences across the data sets, include Kirschbaum and Knafl's (1996) secondary study in which they re-used data from their respective primary studies of family response to a child's chronic illness and parental response to a child's life-threatening illness. In this case, a sub-set of the data from one study and all the data from the other study were combined in the secondary study. The selection of this sample was 'purposive', designed to reflect the different kinds of illness experiences and interactions with professionals. In Angst and Deatrick's (1996) secondary study, the authors drew on data from their individual studies on the ways in which children with cystic fibrosis and their parents defined and managed the illness, and on family decision-making patterns regarding surgery for adolescents with scoliosis. The involvement of children in decisions about their health care was compared and contrasted across these two groups. Thorne et al. (2000) also carried out a secondary study re-using data from two primary research studies which the authors had separately conducted on patients' experiences of chronic illness. This amplified analysis was also comparative in that the studies involved patients with two chronic conditions and living in different countries. Similarly, in their study of disputed diagnoses, Arksey and Sloper (1999) explored lay perspectives and empowerment in relation to obtaining a diagnosis using data from their respective studies of repetitive strain injury and childhood cancer.

In the other group of amplified analyses, one of the authors made use of two or more data sets from their own *oeuvre*. An example of this approach is provided by Bull and Kane's (1996) secondary study of gaps in discharge planning which drew on data from two studies previously conducted by the first author. Similarly, in a study by Ferrell et al. (1996) the authors re-examined data from four studies conducted by the first author and others to describe the impact of fatigue on the quality of life of patients with cancer. In another study of mechanisms of non-compliance among drug users (Fendrich et al., 1996) the authors drew on data from two longitudinal ethnographic studies with which one of the authors had been involved. Other examples include: a secondary study by Pickens (1999: 233) in which the author re-used data from two of her previous studies on mental illness to examine 'the desire for normalcy' expressed by participants in these projects; a secondary study by Cohen (1995) who re-used data from her previous research to examine the sources, properties, and management of uncertainty common to chronic, life-threatening illnesses; and a secondary study by Thorne (1988) who re-used data from her previous research to examine helpful and unhelpful communications in cancer care.

These examples of the re-use of data sets from one's own *oeuvre* reflect an incremental form of research whereby individual studies, rather than forming a disconnected series of works, instead evolve progressively. This approach is well illustrated by a collection of secondary analyses published by the aforementioned Thorne together with Robinson. They later combined the data from their MA research and published two secondary studies on health care relationships (Robinson and Thorne, 1984; Thorne and Robinson, 1988a). Building on this work, the two authors subsequently obtained funding to do a primary study in 1989. Following on from this body of primary and secondary analyses, they individually then published four amplified/supplementary analyses of these accumulated data sets. Thus, Thorne (1990a, 1990b, 1990c) re-used the data to examine mothers' experiences of chronic illness and the meaning of non-compliance for those with chronic illness, as well as completing a doctoral thesis involving the analysis of secondary and primary sources. Robinson (1993) drew on part of the same material to examine the normalization of life by chronically ill members and their families. Although this study is not self-defined as secondary analysis, it is based on the ongoing analysis of a theme arising from these extant data sets.

Assorted analysis

The final variety of secondary analysis to be distinguished was 'assorted analysis' which involves the use of various qualitative data and methodologies. Here the analysis of pre-existing research data is combined with primary data collection and analysis and/or the analysis of additional naturalistic qualitative material. In such studies the balance of these methodologies may vary, with one predominating over the other(s). Like amplified analysis, assorted analysis can be used to provide additional comparative or collateral evidence using different sources of data. Alternatively, it can be used for 'cross-validation' (Thorne, 1994, 1998) in support of the analysis of other types of data.

Five examples of assorted analysis in which the secondary analysis of qualitative data from research studies was carried out in conjunction with primary research were identified in the health and social care literature. In four of these studies, secondary analysis was the main methodology employed. These included two studies by Hutchinson (1987, 1990), neither of which were defined by the author as using secondary analysis, although they have been cited as such (Hinds et al., 1997). Both secondary studies make use of the same primary data from the author's previous research on unprofessional behaviour among hospital-based nurses. In the first 'descriptive' study, the assorted/supplementary analysis focused on nurses' self-care strategies and involved a secondary analysis of a sub-set of the data as well as an additional 20 interviews with nurses to augment the original data and to check initial observations. In the second 'grounded theory' study, the assorted/supplementary analysis focused on how nurses bend the rules for the sake of the patient. Again a sub-set of the data was re-used and an additional 21 interviews were conducted in order to 'verify' the initial observations. In the remaining studies, Doherty et al. (1986) re-used a sub-set of data from an initial 'follow up' study of couples' reactions to marriage encounter weekends; interviews with 50 couples from the original study were also carried out and analysed alongside the pre-existing data (open-ended responses to questionnaires). Finally, Thorne (1990b) re-used data from multiple data sets to explore an emergent theme – non-compliance with professional advice in chronic illness – and also conducted additional interviews with informants for this purpose.

An additional study combining secondary analysis of qualitative research data with primary research was identified in which the secondary analysis was subsidiary to the primary research. Sandelowski and Jones (1996) undertook a primary study of couples' experiences of pre-natal diagnosis and, in particular, their responses to the acquisition of foreknowledge of fetal impairment. The analysis is mainly based on primary source data, comprised of interviews with fifteen women and twelve of their male partners. In addition, 'for purposes of further comparison and validation' (Sandelowski and Jones, 1996: 85) the authors also used information from secondary research data on parenting medically fragile infants collected by another investigative team. This involved reading relevant extracts from ten interviews with parents who had no foreknowledge of their babies' impairments, and four interviews with couples who did have foreknowledge. The information from the secondary sources was used to 'verify our interpretations from the primary source data' (Sandelowski and Jones, 1996: 86).

Finally, four studies were identified which combined the secondary analysis of pre-existing research data with the analysis of naturalistic qualitative material. In two of these studies, the authors examined autobiographies (Cohen, 1995; Thorne, 1988). Another included the analysis of historical documents (Aita and Crabtree, 2000) and another incorporated an analysis of the obstetric literature for physicians' perspectives on ultrasonography, together with the wider literature on other techniques for observing the body, such as x-rays (Sandelowski, 1994b).

Conclusion

In this chapter I have described the various ways in which secondary analysis has been used as a qualitative methodology in social research, based on a review of over 60 studies. A limitation of the review was that it focused mainly on examples of qualitative secondary studies drawn from the health and social care literature (English language publications), although this is an area where most of the pioneers of the methodology work. The identification of studies was also made difficult by the variable ways in which the methodology of the studies was defined. It is therefore likely that some relevant studies were not identified by the search strategies employed. The classification and appraisal of some of the studies included in the review was also limited by the poor accounting of how they were conducted in reports of the research. For example, in some cases, lack of information on the primary study made it difficult to distinguish the respective purposes of the primary and secondary research.

Despite these limitations, the review provided a clearer picture of the ways in which pre-existing qualitative data have been used in social research. In particular, it has shown that the secondary analysis of this material has been mainly pioneered in North America since the 1990s. So far, the methodology has been used to investigate new and additional research questions rather than to verify the findings of previous studies; it has involved the use of single, multiple and mixed qualitative data sets; and it has been conducted mainly using researchers' own, previously collected data and through informal data sharing rather than using archived data sets. In addition, examples of five different types of qualitative secondary analysis have been discerned. The most common type was supplementary analysis, followed by amplified analysis, supra analysis and assorted analysis. Re-analysis was found to be rare in the health and social care literature and in other areas of social research.

These findings would seem to confirm that, compared to the secondary analysis of quantitative data, the methodology is less established in relation to qualitative data sets. They also indicate that the nature of qualitative secondary analysis is somewhat different to its quantitative counterpart in terms of the purpose it serves, the sources of secondary data drawn upon, and the various forms that it takes.[8] These apparent variations in the nature of qualitative and quantitative secondary analysis, as well as in the different types of qualitative secondary analysis, raise a number of epistemological, ethical and methodological issues which will be examined in subsequent chapters.

Notes

1. While the review focuses mainly on the international health and social care literature published in English, some examples of secondary studies published in other areas of social research are occasionally referred to for illustrative purposes.

2. References to these studies are provided in the subsequent sections of the present chapter where different varieties of secondary analysis and examples of these types of studies are described.

3. Two studies were based entirely on archived oral history interviews. These were included although, as we saw in Chapter 1, this type of data are variously defined as 'naturalistic' or 'non-naturalistic' in social research and the emphasis of this book is mainly on the re-use of the latter types of qualitative data.

4. Furthermore, almost all of the secondary studies were designed to investigate further substantive issues, apart from two studies which re-used qualitative data for illustrative purposes in what I will refer to as 'demonstration' projects.

5. I have previously referred to 'assorted analysis' as 'complementary analysis' but now prefer the former term (Heaton, 2000).

6. An equivalent approach was previously highlighted (but unnamed) by Heaton (1998).

7. Other examples of this type of secondary analysis in which part of the original data set is re-used were identified (see Cohen, 1995; Doherty et al., 1986; Gregory and Longman, 1992; Hutchinson, 1987, 1990; Rivard et al., 1999; Sandelowski, 1994b; Tishelman and Sachs, 1998).

8. However, comparisons are limited by the lack of up to date evidence on how quantitative data sets have actually been re-used in secondary research. In particular, while records of usage of some archived data sets are available (see Boddy, 2001), there have been no surveys of practice in quantitative secondary analysis since Hyman's (1972) work on the re-use of survey data. Thus, it is not known to what extent quantitative data are used for purposes of new research or for purposes of verification, nor what sources of data and types of secondary analysis are preferred.

4

Epistemological Issues

The nature of qualitative research	54
The problem of data 'fit'	57
The problem of not having 'been there'	60
The problem of verification	65
Conclusion	71
Notes	71

The extension of secondary analysis to qualitative data raises a number of issues about the compatibility of the methodology with the epistemological foundations of qualitative research. In this chapter following an overview of the main characteristics of qualitative research I examine three key issues that lie at the heart of the debate over whether qualitative data can be re-used in the same ways as quantitative data. These are: first, the problem of data 'fit' – whether or not pre-existing qualitative research data can be legitimately used for purposes other than those for which they were originally collected; second, the problem of not having 'been there' – how the relatively distant relationship of the secondary analyst to the data affects his or her interpretation of the material; and third, the problem of verification – whether or not qualitative data can be re-used for the purposes of confirming, refuting or revising a previous analysis of the same data set. In each case I outline the nature of the problem and, drawing on ideas and evidence from existing work, examine how it has been addressed in theory and in practice.

The nature of qualitative research

Qualitative research is a broad church. It encompasses many traditions of social inquiry dating from the nineteenth century (see Bogdan and Biklen, 1992; Denzin and Lincoln, 2000; Jupp and Norris, 1993;

Murphy et al., 1998). These traditions include different 'schools' (such as the Chicago school), theoretical perspectives (such as phenonemology, symbolic interactionism, naturalism), research protocols (such as grounded theory, framework analysis), methods of data analysis (such as narrative, discourse, content analysis), and types of qualitative data studied (naturalistic and non-naturalistic). While qualitative inquiry does take many forms, there are a number of characteristics which are common to much of this research (Bogdan and Biklen, 1992; Bryman, 1988; Hammersley, 1997b; Murphy et al., 1998). Bryman (1988) highlights six key features:

- the emphasis on 'seeing through the eyes of' the people being studied;
- the description of the social setting being investigated;
- the examination of social behaviour and events in their historical and social context;
- the examination of the process by which social life is accomplished (rather than the end-products or outcomes of interaction);
- the adoption of a flexible and unstructured approach to social inquiry, allowing researchers to modify and adapt their approach as need be in the course of the research;
- the reliance on theories and concepts that have been derived from the data (rather than defined in advance).

While these characteristics generally distinguish qualitative from quantitative research, in practice, there are variations across the different traditions of qualitative inquiry, some of which have features in common with quantitative research (Bryman, 1988).

Although qualitative and quantitative research are generally regarded as distinct modes of inquiry in social research, there are different views on the relationship between, and compatibility of, these approaches. Bryman (1988) distinguishes two broad 'epistemological' and 'technical' perspectives which are similar to the 'paradigm' and 'methodological eclecticism' positions outlined by Hammersley (1997b). In the first, 'epistemological' or 'paradigm' position, quantitative and qualitative research are viewed as separate strategies that are based on incommensurable philosophical assumptions about the nature of reality and the nature of knowledge. Thus, quantitative research is generally associated with positivist and realist epistemologies, upon which the natural sciences are also based. These assume that the social world, like the natural world, has an independent existence which can be observed through scientific methods (although positivists and realists differ as to the extent to which these observations are treated as a reflection of reality or an approximation of it). In contrast, qualitative research is associated with different intellectual positions, such as 'phenomenology', 'symbolic interactionism' and 'naturalism' (Bryman, 1988).[1]

Table 4.1 Axioms of positivist and naturalistic inquiry

Axioms about	Positivist paradigm	Naturalist paradigm
The nature of reality	Reality is single, tangible, and fragmentable	Realities are multiple, constructed, and holistic
The relationship of the knower to the known	Knower and known are independent, a dualism	Knower and known are interactive, inseparable
The possibility of generalization	Time- and context-free generalizations (nomothetic statements) are possible	Only time- and context-bound working hypotheses (idiographic statements) are possible
The possibility of causal linkages	There are real causes, temporally precedent to or simultaneous with their effects	All entities are in a state of mutual simultaneous shaping, so that it is impossible to distinguish causes from effects
The role of values	Inquiry is value-free	Inquiry is value-bound

Source: Reprinted from Lincoln and Guba (1985: 37) with permission of Sage Publications.

These assume that there is no single reality but multiple realities which are contingent on intersubjective understanding. Qualitative methods are used to describe and examine the nature of subjects' perceptions of the world and how their views are constructed, rather than to establish whether or not the views are a true reflection of reality. Accordingly, the science of qualitative research essentially involves demonstrating fidelity to subjects' perspectives rather than correspondence with an underlying 'reality'. The differences in these epistemological positions are nicely illustrated by the example of 'naturalistic inquiry' which was developed by Lincoln and Guba as an alternative to the positivistic paradigm (see Table 4.1).

By contrast, in the second, 'technical' or 'methodological eclecticism', position, qualitative and quantitative research are viewed as complementary rather than opposing research strategies. According to this view, these approaches have different strengths and weaknesses and are therefore suited to the investigation of different kinds of research questions. The decision as to which strategy to use depends on the nature of the question rather than adherence to one epistemology or another. Indeed, qualitative and quantitative methods may even be used in combination in single studies if this is adjudged to be an appropriate means of researching a given problem. Thus, in this position, emphasis is placed on the practical rather than the philosophical aspects of social inquiry.

Both Bryman (1988) and Hammersley (1997b) observe that there are problems with these two positions. For example, the 'epistemological' or 'paradigm' view may be criticized for exaggerating the differences between qualitative and quantitative inquiry, and for failing to reflect the diversity of research within each tradition. As studies of the history and practice of social research have shown different qualitative research traditions have been influenced by various epistemological positions, including positivism (see Denzin and Lincoln, 2000) and realism (see Miller, 2000). Some leading exponents of qualitative research methods have also defined themselves as 'realists' (Miles and Huberman, 1994) or have been lambasted for their 'positivist' disposition (Whyte, 1993: appendix A; 1997). In addition, although the 'technical' or 'methodological eclecticism' position recognizes the value of combining qualitative and quantitative methods, Hammersley (1997b) suggests that there are nevertheless dangers in 'rapprochement', such as losing some of the distinctive features of qualitative research, and failing to give consideration to important theoretical issues through being overly pragmatic. Hammersley also argues that this position still views qualitative and quantitative research as essentially distinct rather than unified strategies for studying the social world. He suggests that a third, more integrated view of qualitative and quantitative research, may be a better alternative to the 'paradigm' and 'methodological eclecticism' positions.

Bryman (1988) suggests that, in practice, researchers tend to 'oscillate' between these views in their work. As I show below, this is certainly the case in the debate over qualitative secondary analysis. Collectively, exponents and proponents of the methodology tend to exemplify the 'technical' or 'methodological eclecticism' position, while sceptics lean toward the alternative 'epistemological' or 'paradigm' position, querying the more fundamental aspects of the process of re-using qualitative data. Their respective views are described in the following, more detailed, examination of the problems of data 'fit', not having 'being there' and verification.

The problem of data 'fit'

In secondary research, the problem of data 'fit' arises because it involves, by definition, the use of data that were originally collected for other primary purposes, or for omnibus secondary research purposes. Thus, in both quantitative and qualitative secondary analysis, issues such as the composition of the sample and the extent of vital missing data may limit the potential for re-using the data for secondary purposes. However, the problem of data fit is compounded in qualitative secondary analysis because of the flexible nature of qualitative research and the character of associated data sets.

Unlike quantitative data sets which generally contain highly structured information on a fixed range of topics, qualitative research tends to produce data sets that are relatively unstructured, rich and diversified. For example, the analysis of qualitative data often begins during fieldwork and informs the remaining data collection. This process of 'sequential' analyses (Miles and Huberman, 1994) allows researchers to leave open the analytical focus of the research and refine it as they proceed. Researchers can then determine which topics to investigate in depth, and how much information to collect, based on their growing knowledge of the field. Similarly, in exploratory studies, open and semi-structured interviews are often used in order to give informants the opportunity to highlight topics related to the inquiry, rather than have these entirely pre-defined by the researchers. Probes may also be used at the researchers' discretion to explore some topics in depth, depending on the informants' responses, and their characteristics and circumstances. Thus, the depth and uniformity of coverage of topics can be variable in qualitative research, depending on how the focus of the primary research was defined and which methods were used to collect the data.

It has been suggested that the extent of data 'fit' in qualitative secondary analysis will depend on three considerations (Hinds et al., 1997; Thorne, 1994). One is the extent of missing data, which may be related to the type of design which produced the data set. For example, in grounded theory, where data collection is an iterative process, shaped by an emerging analytical framework based on ongoing analysis of the data (Glaser and Strauss, 1967) it is likely that some topics will be covered in more depth than others (Hinds et al., 1997). Conversely, one might add that studies based on the 'framework' approach, where data collection is designed to address a particular set of pre-defined aims and objectives (Ritchie and Spencer, 1995) is likely to result in a more even coverage of topics.[2] A second consideration is the degree of convergence between the questions posed by the secondary and original research respectively. That is, the greater the divergence in the aims of the primary and secondary studies, the more likely it is that the data set will not in itself be sufficient for the latter investigation (Hinds et al., 1997). Finally, the fit between the data and the secondary research question may also depend on the methods used to produce the data. For example, comparative secondary studies may only be possible where data sets are comprised of similar types of qualitative data that can be subject to the same methods of analysis (Thorne, 1994).

In the studies reviewed the 'fit' between the data and the research question appears to have been accomplished by the following means. First, in several cases, the studies were designed to address questions that researchers had endogenously derived from their previous analysis of the data set and hence were already grounded in the material. However, as we saw in Chapter 3, some of the questions examined in the secondary studies were more closely related to those of the aims of the original primary work than others. In theory, the issue of data fit is more likely to limit the scope for supra analysis, where the aims of the primary and secondary research are particularly divergent. Conversely, it is less likely to be an issue in supplementary analysis, where the focus of the secondary inquiry is on matters which are, by definition, closer to those investigated in the original work, although the extent and depth of coverage of the relevant topic(s) may be variable. In addition, as we saw in the last chapter, there may be a reverse issue of too close a fit in supplementary analysis, particularly where the same researcher who conducted the original research is re-using the entire data set to explore a research question which is closely related to the primary research. In these cases, the boundaries of secondary analysis and primary research may be difficult to distinguish.

Second, the review provided evidence that, far from treating qualitative data as 'given', researchers had frequently re-shaped it to fit the purposes of the secondary study. For example, they often re-used part of the data set rather than all of it. The selection of these data is perhaps best described as a form of 'sorting' rather than 'sampling', the latter term being reserved for the description of strategies for collecting data in the first place. Thus, sorting was used to segregate sub-samples of the study population, shifting the focus of the secondary analysis to a particular group of informants (see Fendrich et al., 1996; Kearney et al., 1994a; McLaughlin and Ritchie, 1994) and to selectively limit the analysis to certain topics or themes (see Ferrell et al., 1996; Gallo and Knafl, 1998; Gregory and Longman, 1992; Jairath, 1999; Jenny and Logan, 1996). While sorting is not necessarily particular to secondary research, nevertheless it is a technique which secondary researchers often used to manipulate and shape the data set(s) so that it fitted the purpose of their analysis (see also Chapter 6).

Third, in addition to sorting the data, other researchers had augmented the secondary data set with additional primary or naturalistic data, as in the examples of assorted analysis described in Chapter 3. This practice is exemplified by the work of Hutchinson (1987, 1990) who conducted two secondary studies using a data set from her primary research on unprofessional behaviour among hospital-based nurses. In each of these studies, additional interviews were carried out with, respectively, a convenience sample of 20 nurses and a purposive sample of 21 nurses, for the purposes of investigating nurses' self-care strategies and their rule-bending behaviours in more depth.

It would appear then that researchers have dealt with the problem of data fit in practice by exploring questions that were derived from their previous analysis of the same data set, and by re-shaping data sets so that they matched the aims of their secondary work rather than carrying out another primary research study. Indeed, secondary analysis appears to have been utilized as an extension of primary research, rather than as an alternative to it. Researchers' use of pre-existing research data to follow up issues that emerged from their primary work arguably epitomizes the flexible nature of qualitative inquiry and the pragmatic approach of many exponents and proponents of qualitative secondary analysis.

The problem of not having 'been there'

In addition to being flexible, qualitative research is characterized by an emphasis on 'seeing through the eyes of' the people being studied, and on describing and examining the process of everyday life in its historical and social context (Bryman, 1988). To these ends, qualitative researchers typically engage in 'intense and/or prolonged contact' with a 'field' or 'life situation' in order to capture data on subjects' perceptions 'from the inside' (Miles and Huberman, 1994: 6). They also endeavour to stay close to the data by, for example, transcribing their own interviews and immersing themselves in data analysis. This is complemented by drawing on their related research experiences with similar groups, having contact with other members of the same population on study advisory groups and other forums, and by steeping themselves in the relevant literature. As Dale et al. observe, this approach is different to quantitative research in that:

> the [qualitative] researcher becomes the research instrument, with all the 'results' being filtered through her perception and understanding of the social situation in which she is working. The data produced by this means is not only highly dependent upon the researcher but cannot be distanced from her in the way that survey data can. (1988: 15)

Because of this character of qualitative research, they suggest that it is questionable:

> whether it is possible for another researcher to use an interview transcript to reconstruct an in-depth interview without first-hand knowledge of the context of the interview, or whether another researcher can re-analyze the field-notes of an unknown colleague.

> Research of this kind sets out to capture an understanding of a process through a methodology that immerses the researcher in the chosen topic. In these circumstances, it seems unlikely that the re-analysis of either interview transcripts or field-notes by an outsider could give more than a partial understanding of the research issues. (Dale et al., 1988: 15)

This problem of not having 'been there' has been identified as a key concern among qualitative researchers in a survey by Qualidata (Corti, 1999, 2000; Thompson, 2000a). While it has generally been discussed in the context of re-using other researchers' data (see Corti, 1999, 2000; Corti and Thompson, 1998; Fielding and Fielding, 2000; Hammersley, 1997a; Hinds et al., 1997; Thompson, 2000a; Thorne, 1994), as I show below, similar issues apply to the re-use of researchers' own data sets.

Re-using other researchers' data

Very few of the secondary studies in the review were based entirely on the re-use of other researchers' data. In those that were, no particular comments were made on the choice and experience of using this source of data, with one major exception. This was a study by Weaver and Atkinson (1994) who re-used Roth's fieldnotes from his work on social relations in tuberculosis sanitaria which he had made available to Atkinson. The data were re-used for the purposes of developing and illustrating a book on micro-computing strategies in qualitative research rather than for a 'comprehensive' analysis of a particular question (Weaver and Atkinson, 1994: 3). They opted to use this data, rather than any of their own, because:

> we believed that it would be of considerable benefit to distance ourselves from the data and the outcome of the analysis. Approaching the data 'cold' was likely to free us to look dispassionately at the processes of data searching and analysis rather than at specific outputs. (Weaver and Atkinson, 1994: 2)

This was confirmed by their experience, the use of this source of data enabling them to 'concentrate on the micro-computing strategies themselves, and to subordinate the substantive issues' (Weaver and Atkinson, 1994: 3). However, they also note that while their use of 'cold' data for this purpose was 'liberating', their 'distance' from the data was also a limitation in some respects. As they explain (and I quote at length because it illustrates the issues so well):

> When one reads fieldnotes or transcripts, tacit knowledge is inevitably brought to bear. In most ethnographic projects the analyst draws on his or her personal experience of 'the field'. Reading and re-reading the fieldnotes is an active, interpretive process. One reads knowledge and

experience into the notes and the other data. In the course of this project, on the other hand, we were relying on 'the data' alone, which as a consequence often felt 'empty' or 'thin'. Because we were working in this way, we were not able to evaluate and exemplify the procedures whereby data may be coded and analysed concurrently with fieldwork. Since we were not collecting the data, we were obviously not able to develop a fully 'grounded' approach whereby our emerging categories and theories could inform further data collection. In the course of real research projects, the close linkage between data analysis and fieldwork is important; here we cannot address that issue explicitly. (Weaver and Atkinson, 1994: 3–4)[3]

More generally, some of the difficulties of re-using other researchers' data have been touched on in Sanjek's (1990a) fascinating book on the use of fieldnotes in social research. In it anthropologists and ethnographers describe the difficulties of completing the work of colleagues who have died based on their fieldnotes. As Ottenberg (1990) explains, anthropologists record their observations of a culture using fieldnotes which may have been reconstructed from contemporaneous 'scratch notes' made in the field. In addition to these records there are what Ottenberg calls 'headnotes' – memories of the field research, some of which may also be written down in research diaries. This tacit knowledge also informs the researcher's analysis. According to Ottenberg:

> headnotes are always more important than the written notes. Only after the author is dead do written notes become primary, for then the headnotes are gone. Headnotes are the driving force, albeit subject to correction by the fieldnotes. The written notes have a sacred quality that is also an illusion. The process of employing fieldnotes should make them an adjunct to the more primary headnotes, which lead the written form, even though for living anthropologists, writing up headnotes without written notes – as when the latter are lost – presents immense difficulties. Only a few have attempted it. For most of us, both are required. (Ottenberg, 1990: 147)

The importance of 'headnotes' is evident in Sanjek's (1990b) account of how Barth found it frustrating trying to complete Pehrson's work on the Marri Baluch nomads of Pakistan on the basis of extensive fieldnotes and letters following the latter's death in 1955. Barth was only able to do this after he collected additional primary data he needed in order to complete the analysis.

However, as Ottenberg (1990: 156) points out, there has been an epistemological shift in ethnography from positivism to humanism and in the related treatment of fieldnotes as 'scientific data' and 'interpretive texts'. According to the latter perspective, the interpretation of data is bound to vary from analyst to analyst, and when re-visited by a researcher in the light of his or her changed biographical circumstances. Or as Bond more eloquently puts it:

> fieldnotes simulate and are part of human experiences. The notes are thus living, mutable texts; they are a form of discourse whose content is subject to constant re-creation, renewal, and interpretation. The immutable documents and the mutable experiences stand in dialectical relationship, denying the possibility of a single reality or interpretation. (1990: 274)

Viewed in this way, the use of other researchers' data presents similar interpretive challenges to those faced by researchers re-using their own data, or researchers who carry out primary qualitative research in teams, sharing data as part of this process (Heaton, 1998).[4] In each case, the respective contexts of data collection and (re)analysis have to be taken into account.

In response to the general issue of how researchers can make sense of material collected by other researchers, proponents of qualitative secondary analysis have suggested ways in which primary and secondary researchers can bridge the gap in their relationship to the data. In particular, it has been suggested that, when depositing their data set with an archive, primary researchers should supply relevant study documentation and contextual information to facilitate secondary analysis of the material (Corti and Thompson, 1998; Fink, 2000). This 'meta-data' includes fieldnotes, research diaries, notes and correspondence relating to the conduct of the primary research. It has also been suggested that, where possible, secondary analysts could contact the primary researchers who collected the data to discuss the original work and become sensitized to the context of the primary study (Hinds et al., 1997; Szabo and Strang, 1997).[5] While it was not clear to what extent analysts had drawn on any meta-documentation in the studies reviewed, it was reported that primary researchers were consulted in some of the projects (Clarke-Steffen, 1998; Fielding and Fielding, 2000; Szabo and Strang, 1997, 1999).

Re-using auto-data

The fact that the vast majority of the studies reviewed were based on data derived from primary studies that the analysts had themselves been involved in may indicate that this is perceived to be less difficult and problematic compared to the practice of re-using other researchers' data.[6] The advantages of re-using auto-data were made very explicit by West and Oldfather (1995) in their outline of a type of qualitative secondary analysis called 'pooled case comparison'. They argue that:

> An important difference between the use of a data bank and our method [pooled case comparison] is that in analysing their own data, researchers have the advantage of deeper knowledge of the contexts from which the data were derived. In the studies included in our pooled case comparison, we were participant observers. Thus, as we conducted the analysis, we brought to the process a level of background knowledge unavailable to those working from a bank of others' data. Reading a research report, or even examining raw data, is not the same as being present in a research context... Direct engagement in interpretive research brings about a different quality of knowing. This 'participatory knowing' cannot be achieved through the eyes of even the most interested researcher who was not bodily present in the research setting. (West and Oldfather, 1995: 456)

While the re-use of one's own data appears to confer the advantage of being close to the data, this cannot be taken for granted. As I have previously suggested (Heaton, 1998), where primary research is carried out by a team of researchers (rather than a solitary researcher) individual members of the team may have varied involvement in the research design, fieldwork, data processing and analysis and hence differential knowledge of the resulting data set. Accordingly, the fact that the analyst was involved in the primary research does not in itself mean that he or she is any better placed to perform a secondary analysis than an independent researcher. Indeed, primary researchers working in teams arguably face similar issues of interpretation to secondary analysts in terms of having to make sense of data that were collected by other team members (Heaton, 1998). This issue also applies to teams of secondary analysts who, as in pooled case comparison, combine their data from studies that were independently carried out.

Some researchers' experiences also suggest that, even where individual researchers re-use only their own data, there are related issues about them not having 'been there' – for a while. For example, in re-reading his own fieldnotes, Atkinson (1992) found that these were 'evocative' and triggered memories of the field, some of which were stronger than others. He observes that:

> After the passage of time, the notes are alien in some respects. One comes to them 'cold'. Reading them is, however, different from the reported experience of working with someone else's field data (Lutkehaus, 1990) – for they can still evoke a lived experience. (Atkinson, 1992: 460)

Mauthner et al. (1998) also discuss how their analysis of data which they collected some time ago was affected by their waning memory of the fieldwork and by changes in their social status which shaped their interpretation of the data. They express concern that secondary data might be used by empiricist qualitative researchers who do not give due consideration to the ways in which knowledge about the world is context-bound. For these reasons, they caution against crude or naive realist forms of secondary analysis in which the data are treated as given and the contingent nature of secondary analysis is ignored. They also imply that qualitative data collection and analysis are best conducted as contiguous processes, at the height of a researcher's involvement with the research.

In sum, it would seem that the problem of not having 'been there' is complex and exposes some of the epistemological tensions in qualitative research. On the one hand, there is a view among more sceptical researchers, such as Mauthner et al. (1998), that qualitative data are best analysed by the researcher(s) who were involved in collecting them, at or close to the time that they were collected. According to this view, which is consistent with the 'epistemological' or 'paradigm' position outlined above, no amount of meta-documentation or consultation with primary researchers could ever compensate for not having 'been there'. On the other hand, there is a another view, held by proponents of qualitative secondary analysis, that having recourse to meta-data and to primary researchers can help independent researchers to interpret data that they did not collect, or which they collected sometime ago. This perspective, which is closer to the alternative 'technical' and 'epistemological eclecticism' view of qualitative research, is exemplified by Fielding and Fielding who justified their secondary analysis of an independent data set on the grounds that:

> Researchers are used to tracing the mediating effects of reflexivity in primary data analysis and we believed that the recovery of contextual features in secondary data analysis is a practical rather than an epistemological matter. (2000: 679)

Thus, for these researchers and archivists, the analysis of one's own and other researchers' data, and the quality of the data as a secondary resource, raises different interpretive and technical challenges, but does not preclude the possibility of secondary as well as primary analysis of qualitative data sets.

The problem of verification

While it is generally accepted that findings of quantitative research can be verified through the re-analysis of data sets by independent researchers, in qualitative research there are radically different views on this issue. Some traditions of qualitative inquiry, such as conversation analysis, were founded on the positivistic principles of the natural sciences and, as I show below, this is reflected in the ways in which raw data are treated in their work. Other traditions, such as naturalistic inquiry (Lincoln and Guba, 1985), were developed as an alternative to positivism and have different concepts and strategies for establishing the 'trustworthiness' of studies – including 'triangulation', 'negative case analysis', 'member checks', 'audit trails' and 'referential adequacy' – some of which have been adopted more widely in qualitative research. The concepts of the 'audit trail' and 'referential adequacy' are of particular interest here because they both involve the retention of raw data in order that other researchers can independently check the authenticity of the work.[7] These various approaches to external verification, and the extent to which they featured in the secondary studies reviewed, are examined in more detail below.

Publication of 'raw' data

In reports of quantitative research, aggregate data are usually presented in tables or appendices with the text. As we saw in Chapter 2, there have been periodic calls for researchers and editors to publish more raw statistical data with research reports in order to facilitate the independent scrutiny of the work. In qualitative research, however, there are different conventions for rendering and representing textual and graphic data in final reports.

For example, language is commonly used as data in qualitative research (Miles and Huberman, 1994). In interviews and focus groups, the words of informants are often recorded on audio tape and transcribed for purposes of analysis. These recordings may be fully transcribed 'verbatim' or selectively transcribed (when, for example, the theme(s) to be analysed are confined to sections of the tapes). The resulting transcripts may also be accompanied by the researcher's commentary on the circumstances of the interview and any relevant observations and off-tape discussions. Alternatively, interviews and observational work may be recorded manually. These notes may include occasional quotes and paraphrased remarks as well as descriptions of the encounter and social setting of the interaction.[8] Once rendered into textual form, the data are then subject to analysis. Increasingly, the textual data are input onto a computer and analysed using programmes such as Atlas.Ti and NVivo which were developed to assist with the analysis of this type of data. Where researchers have data in different formats, they may develop their analyses based on listening to or watching the tapes, as well as reading the transcripts and/or e-transcripts, or they may prefer to work mainly with one of these formats. Their published analyses also vary in terms of how they present these data in reports of their work. In most reports, quotations are selectively used to illustrate the analysis and often only represent a small portion of the overall data upon which the analysis was based. Quotations may also be edited to make them read more smoothly and to disguise the informant's identity.

In some forms of qualitative research, however, a very different approach is adopted. In conversation analysis, for example, audio or video recordings of naturally occurring interaction are rendered using special transcription systems and symbols. These data form the sole basis of the analysis and no other contextual information is provided, the focus being on the underlying rules of social interaction. Published analyses are usually accompanied by extracts of the transcribed data, upon which the analysis is usually founded (rather than the tapes – but see below). The assumption underlying this convention is that the analyst's observations can be checked against the data. This was seen by Sacks (who pioneered the methodology) as a strength of conversation analysis, compared to other forms of qualitative inquiry of the time:

> The difference between [ethnography] and what I'm trying to do is, I'm trying to develop a sociology where the reader has as much information as the author, and can reproduce the analysis. If you ever read a biological paper it will say, for example, 'I used such-and-such which I bought at Joe's drugstore'. And they tell you just what they do, and you can pick it up and see whether it holds. You can re-do the observations. Here, I'm showing my materials and others can analyze them as well, and it's much more concrete than the Chicago stuff tended to be. (Sacks, 1992: vol. 1: 27. Quoted in Hutchby and Woofitt, 1998: 26)

Although conversation analysts still generally publish extracts of transcripts alongside their analyses, as Ashmore and Reed (2000) have shown, there are different views within conversation analysis as to whether analyses can be based on transcripts alone or whether the tapes also need to be referred to. Thus, some believe that transcripts effectively capture the data while others believe that they are too crude a device for analytical purposes. The implication of this debate is that in order to verify an analysis, access to the tapes as well as to the transcribed data may be required.

Lapadat (2000) has argued that variations in the use of transcripts in qualitative research reflect differences in the epistemological assumptions about the nature of this data. Thus, in early linguistics, the transcript was assumed to be a 'one-to-one match with the spoken words' and the 'sum of the observable event' (Lapadat, 2000: 207). From this positivistic perspective, it is assumed that language can be captured and observed through the transparent medium of the transcript. This also means that transcripts can be used by other researchers to check the veracity of the analysis, as is illustrated by the work of conversation analysts. However, in other forms of qualitative research, transcripts are regarded as a re-presentation of the interaction, re-creating the interaction referred to. From this interpretivist standpoint, transcripts shape and construct the interaction they depict and cannot be used unproblematically in social research.

Although conversation analysts do generally include relevant extracts of the 'raw' data in their publications, none of the studies reviewed involved the re-use of this source of data for the purposes of confirming, refuting or revising the findings of previous research.[9] Nor did any of the other secondary studies reviewed involve the re-use of qualitative data contained in publications for the purposes of verification, although as noted in Chapter 4, some classic re-studies have involved the review of evidence in reports (rather than the 'raw' data itself). Some researchers also drew on published 'naturalistic' data such as autobiographies and other documentary material for the purposes of confirming their secondary analysis of non-naturalistic research data (Cohen, 1995; Thorne, 1988; Thorne et al., 1999), and a research publication was used as secondary data in one study exploring images of death and attitudes towards dying (Pascalev, 1996).

Audit trails

The 'audit trail' was proposed by Guba (1981) as a way of allowing external auditors to check the 'dependability' and 'confirmability' of naturalistic inquiry and has been described as possibly the 'single most important trustworthiness technique available to the naturalist' (Lincoln and Guba, 1985: 283). Based on the fiscal audit model, an audit trail is basically a record of the research process which primary researchers keep in order to allow auditors to check that the records reflect the work undertaken (dependability) and that the findings of the research are supported by the data (confirmability).

Drawing on the principles and procedures of the audit trail outlined by Halpern (1983), Lincoln and Guba (1985) state that the audit trail includes all documents relating to the study (such as the research protocol, study correspondence, research tools, and methodological notes) as well as the raw data and associated analyses. Following Halpern, they divide the audit procedure itself into five stages. In the 'preentry' stage a potential auditor is identified and an agreement formed as to the requirements of the audit inquiry. In the second stage, the auditor reviews the records at a time agreed with the primary researcher. At this stage, revisions to the audit trail may be made if the study is still ongoing or the auditor may decline to continue if his/her conditions have not been met. In the third stage, the parties agree a binding contract to proceed with the audit. They also agree the form that this will take and their respective contributions to it. The fourth phase is concerned with assessing the confirmability and dependability of the inquiry. In the fifth and final stage, the auditor prepares a report or 'letter of attestation' outlining the findings of the audit. According to Lincoln and Guba, these steps can all be completed in a week to ten days.

Hammersley (1997a) has identified a number of problems with the audit model. One is that it implies that qualitative research findings can be validated by keeping an audit trail. Although Lincoln and Guba recognize that there is no 'unassailable' truth, at the same time they seem to promote the audit trail as a method which qualitative researchers can use to persuade others of the 'trustworthiness' of their work. In other words, there is an implicit commitment to some kind of hierarchy of knowledge whereby some accounts are better – or more trustworthy – than others. Another problem is that the audit trail can never completely reproduce the research process and render it transparent to others.[10] As Hammersley points out, the concept of the audit trail rests on a fallacious assumption that social research is carried out in a neat and rational fashion when in fact it is a messy and contingent process which is not easily recorded.

Despite further explication of the audit trail (Schwandt and Halpern, 1988), and Lincoln and Guba's (1985: 379) expression of hope that auditors might one day come to be listed in the Yellow Pages, the external auditing of qualitative research is not *de rigueur* in contemporary qualitative research.[11] Although audit trails were kept in some of the secondary studies reviewed (Logan and Jenny, 1990; Szabo and Strang, 1997, 1999; Thorne et al., 1999), it is not clear if these actually culminated in an external audit or were just used for internal quality assurance purposes. However, the issue of verification and, more generally, the quality and accountability of qualitative research has received increasing attention in the literature (see Marshall and Rossman, 1999; Seale, 1999). For example, one of the features of the 'framework' approach often used in social policy research is that 'the analytic process, and the interpretations derived from it, can be viewed and judged by people other than the primary analyst' (Ritchie and Spencer, 1995: 176). The ways in which secondary analysts have accounted for the conduct and quality of their work will be examined further in Chapter 6.

Referential adequacy

Like the audit trail, the concept of 'referential adequacy' was also adopted by Lincoln and Guba (1985) as a way of establishing the trustworthiness of naturalistic inquiry. Drawing on Eisner's (1975) work, they argue that primary researchers could usefully retain a portion of their raw data and archive it to allow the researchers and other 'critics' to access it later for the purposes of testing analyses of the material. They claim that:

> Aside from the obvious value of such materials for demonstrating that different analysts can reach similar conclusions given whatever data categories have emerged – a matter of reliability – they can also be used to test the validity of the conclusions. Sceptics not associated with the inquiry can use such materials to satisfy themselves that the findings and interpretations are meaningful by testing them directly and personally against the archived and still 'raw' data. A more compelling demonstration can hardly be imagined. (Lincoln and Guba, 1985: 313)

However, they do acknowledge that researchers might be reluctant to 'surrender' part of their raw data to an archive and that there are issues about the 'representativeness' of the data selected for archiving. They also add a cautionary note that:

> Naturalists using the referential materials are likely to want to 'peel the onion' to a different layer, demonstrating less interest in the original analyst's findings than in developing their own. For all these reasons the referential adequacy approach does not lend itself well to the more practical-minded or resource poor. Nevertheless, when resources and inclinations permit, the storage of some portion of the raw data in archives for later recall and comparison provides a rare opportunity for demonstrating the credibility of naturalistic data. (Lincoln and Guba, 1985: 314)

It is unclear to what extent, and how, referential materials have been used in qualitative research in general. However, in the studies reviewed, only one referred to the concept. This was a study by Cohen (1995) of perceptions of uncertainty among parents of children with chronic and life-threatening illnesses, where autobiographical accounts were used to check the analysis of the interview data (and these were also included in the analysis). A similar strategy was also used in a study of stress and injury in farming, where data from three focus groups were used to 'validate and reverify the findings' of six other focus groups that were coded and included in the secondary analysis (Kidd et al., 1996: 227). Thus, it would appear that although 'audit trails' and 'referential adequacy' are possible means by which qualitative researchers can establish the 'trustworthiness' of their work, there are doubts about to what extent they are actually used in practice – and also whether they have been used in full, to facilitate the external verification of findings, or in a more limited way, as part of internal efforts to verify the work.

Conclusion

In this chapter, I have suggested that the epistemological problems raised by the development of secondary analysis as a qualitative methodology reflects wider tensions between quantitative research and qualitative research, and between different traditions of qualitative research. Thus, the problems of data 'fit', not 'being there' and verification are all regarded as particularly problematical in qualitative secondary analysis because of the distinctive character of qualitative research and nature of the resulting data sets. However, within qualitative research, there have been different responses to these problems. On the one hand, exponents and proponents of qualitative secondary analysis have generally adopted a practical approach to these problems, exemplifying the 'technical' or 'methodological eclecticism' view of qualitative research and its relationship to quantitative research, while sceptics have drawn attention to the dangers of these tools falling into the wrong (epistemological) hands. At the same time, exponents of qualitative secondary analysis have not shown a great deal of enthusiasm for full-scale re-analyses of previous qualitative studies, nor for fully utilizing techniques such as the audit trail and referential adequacy for establishing the trustworthiness of qualitative work, both of which involve the external scrutiny of the work. Whether this is because these particular methods are perceived to be less compatible with the tenets of qualitative research or because of practical difficulties in implementing them in practice is not clear. However, exponents' use of related methods for establishing the trustworthiness of qualitative work (see Chapter 6 for more on this) suggests that they do not generally regard all such measures as anathema to qualitative research or technically difficult to apply.

It would seem, then, that qualitative secondary analysis has been used and promoted in ways that are broadly consistent with the main tenets of qualitative research. In particular, I have suggested that the methodology epitomizes the flexible character of qualitative research, enabling researchers to find innovative ways of using pre-existing data. I will return to this theme in the final chapter where I discuss the place of qualitative secondary analysis in the contemporary landscape of qualitative research and consider the implications for the future development of the methodology. But first I look at the ethical and legal issues, as well as the related technical matters, arising from the use of the methodology.

Notes

1. The term 'naturalism' has also been used in the positivist paradigm, where it has a different meaning (Bryman, 1988).

2. This approach, which is often used in policy research, is less inductive than most forms of qualitative research (Pope et al., 2000).

3. It may be added that the difficulties Weaver and Atkinson highlight relate, in this case, not only to the re-use of another researcher's data but also to their use of the data for methodological rather than substantive purposes (which may well have amplified the 'distance' between them and the data).

4. For an analysis of the issues associated with doing ethnographic and qualitative research in teams, see Erickson and Stull (1998) and Gates and Hinds (2000).

5. Hinds et al. (1997: 414) suggest that 'Actions that help the [secondary] researchers to feel close to a condition of "having been there" and to imagine the emotions and cognitions experienced by the participants and the researchers during data collection and analysis could be particularly valuable'.

6. Of course, the difficulties of re-using other researchers' data include not only epistemological issues, but practical issues such as the relative lack of availability of qualitative data sets, and difficulties accessing them.

7. In contrast, the other strategies generally involve internal verification and quality checks (rather than external peer assessment by independent researchers). See also the section on 'quality assurance' in Chapter 6.

8. Indeed, in ethnography, the use of fieldnotes is often preferred over the tape-recorder for logistical and other reasons. For example, in their study of therapeutic communities, Bloor et al. (1988) comment:

> Parenthetically, where direct access [to group settings] had been granted, we would not regard audio recordings as an acceptable substitute for fieldnotes in ethnography. Leaving aside the technical problems [of recording]... and [the] logistical problems of transcribing and analyzing many hours of recordings, we agree with Hammersley and Atkinson (1983) that the ethnographer is his or her own research instrument. The tape recorder is a 'cultural dope' and is of only supplementary value for an ethnographer. (1988: 207–8)

9. And as noted in Chapter 1, conversation analysts would not define such studies as 'secondary' analysis anyway as they make no distinction between primary and secondary data, or between primary and secondary research; that is, all researchers are perceived to have an equal relationship with the data regardless of who captured it (provided that it is rendered transparent to others using their special system of notation). For an example of conversation analysis using shared tapes of social interaction see Paget (1983).

10. Indeed, it may be added that the audit trail is an integral part of the original study and the ways in which the data set is constructed. Thus the 'transparency' of the inquiry is a product of the use of the audit trail or equivalent techniques for achieving this effect.

11. The additional time and costs involved (Miles and Huberman, 1994) have probably been a factor in this, as well as the philosophical resistance to the idea that the interpretation of qualitative data can be externally confirmed by the inspection of records.

5
Ethical and Legal Issues

Ethical and legal frameworks for social research	74
Informed consent	77
Confidentiality	81
Copyright	83
Data protection	85
Defamation	85
Conclusion	86
Notes	86

The re-use of data for purposes other than which it was collected – whether this is by the same researchers who were involved in the primary research or by other researchers – raises a number of ethical and legal issues. Should researchers seek and obtain informed consent for secondary research and, if so, when is this best done? How can qualitative data be formally and informally shared between researchers without contravening confidentiality agreements? And what are the implications of laws on the ownership and use of data for qualitative secondary analysis? In this chapter, I examine these and related ethical, legal and professional issues relating to the collection, retention and re-use of qualitative data in social research. I begin by describing the existing ethical and legal frameworks for social research and how they pertain to the conduct of secondary research in particular. The rest of the chapter examines in more depth the key issues of informed consent, confidentiality, copyright, data protection, and defamation from the point of view of researchers who are interested in donating and/or re-using qualitative data sets. These sections draw on existing national and international guidelines and codes of practice (in English) relating to data retention and data sharing in social research and related fields of work (particularly medicine).[1]

Ethical and legal frameworks for social research

The ethical and, to a lesser extent, legal frameworks for social research have been defined and disseminated by four main groups: professional associations in their 'codes of practice' for members of these organizations; funders of social research in their guidelines on good research practice; ethics committees which regulate social research carried out in some contexts (such as the National Health Service in the United Kingdom where all research involving patients is subject to ethical approval, and those universities or university departments which have policies that require research involving human subjects to undergo an ethical review); and other groups, such as Qualidata and individual data archives, which have developed policies and procedures concerning the preservation, documentation and re-use of social research data.

Although these guidelines vary, they are generally informed by fundamental principles previously established in international declarations such as The Nuremberg Code (International Tribunal of Nuremberg 1947 (1990)) and The Declaration of Helsinki (World Medical Association, 1964).[2] Some of the more established guidelines have also been revised and updated over time to reflect changing views on what constitutes good research practice and to incorporate new regulations and laws affecting research work, including national and international legislation on privacy, ownership of data, exploitation of ideas, and the management and use of personal data. However, as I suggest below, existing guidelines vary in the extent to which they provide clear information and advice specifically on the conduct of secondary as opposed to primary research. In addition, they only apply to some social researchers, such as those who follow particular codes of practice, those who are obliged to seek ethical approval in their field of work or by their employers, and those who agree to conform to ethical standards stipulated by data archives.

Several professional associations and learned societies representing social scientists in various countries have developed codes of practice for their members. These include the American Sociological Association (1997), the Canadian Sociology and Anthropology Association (1994), the British Sociological Association (undated, c. 2002), the International Sociological Association (2001), the Oral History Society (undated), the Social Research Association (2002), the Society for Research in Child Development (1990–91), and the Association of Social Anthropologists of the UK and the Commonwealth (1999).[3] While they all include a focus on key issues such as informed consent and maintaining confidentiality, they vary in the extent to which they address these issues in relation to both primary *and* secondary research. In its 'Statement of Ethical

Standards', the American Sociological Association has for some time provided guidelines on the issues of informed consent and confidentiality in relation to the practices of data sharing and data retention (Boruch and Cordray, 1985). Another group which provides relatively detailed information on the legal rights of research participants, as well as advice on data sharing, is the Association of Social Anthropologists of the UK and the Commonwealth (1999). Other groups, such as the Social Research Association (2002) and the British Sociological Association (undated, c. 2002), have also recently revised their guidelines and now provide more advice specifically on the archiving and re-use of quantitative and qualitative data.

A number of research councils and charities that fund social and medical research also issue their own guidelines on good research practice with information and advice on researchers' responsibilities and research participants' rights. These include the United Kingdom's Medical Research Council (2000), The Wellcome Trust (2002), The National Children's Bureau (undated),[4] The Research Council of Norway (The National Committee for Research Ethics in the Social Sciences and the Humanities – The Research Council of Norway, 2001), the National Institutes of Health (1995), the Australian National Health and Medical Research Council (1999), and the Canadian Medical Research Council, National Science and Engineering Research Council and Social Sciences and Humanities Research Council (Medical Research Council of Canada et al., 1998). Again, the main focus of these guidelines is on the conduct of primary research, although some of the medical guidelines in particular include advice on good practice in respect of the retention of data sets by primary researchers (in case results need to be verified).

One major funder of social research in the United Kingdom, the Economic and Social Research Council, commissioned the law firm Allen and Overy to investigate and clarify the legal principles governing the collection, use and ownership of research data in the social sciences. Allen and Overy's (1998) final report covered the laws in England and Wales relating to copyright, confidentiality, defamation and data protection and included a set of draft guidelines for researchers which distinguish the issues relating to quantitative and qualitative data. The Economic and Social Research Council (undated, c. 1998) subsequently produced guidelines based on this report, entitled 'Guidelines on copyright and confidentiality: legal issues for social science researchers' which are of relevance to both primary and secondary researchers involved in quantitative and/or qualitative research.

Another important ethical framework for social researchers is provided by ethics committees and equivalent organizations that review research protocols to ensure, among other things, that the issues of informed consent and confidentiality are properly addressed. In the United Kingdom, these include Local and Multi-centre Research Ethics Committees (LREC and MRECs) in the National Health Service, to which protocols for research involving patients and their families have to be submitted for approval.[5] In the United States and Canada, Institutional Review Boards (IRBs) and Research Ethics Boards (REBs) perform a similar function for research involving human subjects.[6] Many universities in the United Kingdom and other countries also have ethics committees which review protocols as well. These groups vary in the extent to which they explicitly recognize and allow for the possibility of data sets resulting from primary studies being archived and shared in the future, or being re-used by the same researchers who collected them for other purposes. For example, application forms for ethical approval do not always cover this issue directly, and successful applicants are not necessarily provided with clear advice on the circumstances in which data can and cannot be retained and re-used in the future.

Finally, specialist organizations such as Qualidata and various individual data archives have also developed guidelines on the ethical, legal and related practical aspects of depositing and re-using quantitative and qualitative research data in archives. Qualidata's guidelines include a focus on issues relating to the collection and archiving of interview data (Qualidata, undated a; see also Corti et al., 2000), including interviews from studies involving children and people with learning difficulties where special procedures have been developed for gaining their consent to participate in primary research and to other possible uses of the data (Qualidata, undated b; see also Rolph, 2000). Qualidata has also provided advice on the legal issues involved in archiving various types of qualitative data, based on the aforementioned Allen and Overy (1998) report (Qualidata, undated c). In addition, the UK Data Archive and Qualidata have produced practical guidelines on processing qualitative data in preparation for archiving which includes related advice on techniques for maintaining the confidentiality of this type of data (UK Data Archive, undated).

Other archives also provide information and advice on the ethical and legal aspects of their policies and procedures regarding the donation and re-use of data sets. These generally relate to quantitative rather than qualitative material, reflecting the types of data they tend to accept, although some issue guidelines on archiving qualitative data too. For example, the Inter-university Consortium for Political and Social Research's (ICPSR) handbook for researchers on how to prepare data sets for deposit under the National Institutes of Justice (NIJ) research program includes advice on how to preserve confidentiality in quantitative and qualitative data (Marz and Dunn, 2000). Both data donors and secondary data users may be required to sign agreements with data archives confirming that the data were collected or will be used in accordance with the ethical and legal mandates of the archive.

This then is the complex ethical and legal framework in which researchers collect and/or access data for social research. In the rest of the chapter, I will look in more depth at the key issues of informed consent, confidentiality, copyright, data protection and defamation and their implications for qualitative researchers interested in donating and/or re-using qualitative data sets.

Informed consent

The principle of informed consent was outlined at the Nuremberg Trials following the forced participation of concentration camp inmates in medical experiments during World War II (Bulmer, 1979). It requires that research participants are fully informed of the nature of the work and any possible risks to themselves, and that they freely agree to take part. Informed consent is now a fundamental principle of all research involving human subjects, although there are some circumstances in which it may be waived. For example, the guidelines of the American Sociological Association recognize that this might be allowed where the research involves 'no more than minimal risk for research participants' or where the research could not be practicably carried out if informed consent was required (American Sociological Association, 1997: Section 12.01).[7] However, it is noted that waivers of consent 'require approval from institutional review boards or, in the absence of such boards, from another authoritative body with expertise on the ethics of research' (American Sociological Association, 1997: Section 12.01). In addition, despite these various guidelines, informed consent may not always be properly obtained and/or respected, as inquiries into the use of children's organs kept for research purposes after post-mortem examinations at Bristol and Alder Hey hospitals in England revealed.[8]

Is it necessary to obtain informed consent for pre-existing data to be re-used in secondary research? According to the American Sociological Association (1997), researchers may use publicly available information, such as data sets deposited in archives, without obtaining additional consent, although if they have doubts about this they are again advised to consult with an Institutional Review Board or an authority on ethics before proceeding. Similarly, the Social Research Association allows for the secondary use of archived records but reminds researchers of their duty to participants in this statement:

> In cases where subjects are not approached for consent because a social researcher has been granted access, say, to administrative or medical records or other research material for a new or supplementary inquiry, the custodian's permission to use the records should not relieve the researcher from having to consider the likely reactions, sensitivities and interests of the subjects concerned. Where possible and appropriate, subjects could be approached afresh for consent to any new enquiry. (Although this is not required under the UK Data Protection Act as long as there are no additional consequences for the data subject.) (Social Research Association, 2002: Section 4.3(c))

While these and other guidelines cautiously allow for secondary research to be carried out without informed consent, they do not offer more specific advice on the circumstances in which quantitative and qualitative data, whether anonymized or not, may be re-used without consent. Similarly, they do not comment on whether and, if so, when informed consent can be waived by researchers who want to re-use their own data for other purposes, or who want to informally share their data with colleagues.[9] These are significant gaps in existing guidelines, particularly given the sensitive and personal nature of some qualitative work, and the risk that individuals might be identified even after names and details have been disguised.

Some professional associations now advise members that, as part of the process of obtaining consent for participation in primary studies, research participants should be informed about any possibility that the information they provide may be shared with others. For example, the Association of Social Anthropologists of the UK and the Commonwealth states that:

> The long period over which anthropologists make use of their data and the possibility that unforeseen uses or theoretical interest may arise in the future may need to be conveyed to participants, as should any likelihood that the data may be shared (in some form) with other colleagues or be made available to sponsors, funders or other interested parties, or deposited in archives. (Association of Social Anthropologists of the UK and the Commonwealth, 1999: 4 (f) 'Negotiating Informed Consent')

Similarly, the British Sociological Association suggests:

> Where there is a likelihood that data may be shared with other researchers, the potential uses to which the data might be put must be discussed with research participants. (undated, c. 2002: Item b) iii))

At the same time, researchers and archivists have begun to explore ways in which informed consent could be explicitly obtained for secondary studies using retrospective or prospective strategies (see Alderson, 1998; Backhouse, 2002; Corti et al., 2000; Hood-Williams and Harrison, 1998).

Gaining informed consent retrospectively involves going back to the research participants in the primary study and obtaining their consent *de novo* for the secondary analysis.[10] The main advantage of this approach is that participants are directly informed about the nature of the secondary study by the researchers who will be doing the analysis. However, this approach may not always be feasible. For example, participants' names may be withheld from researchers who were not part of the primary study team because of confidentiality agreements, or they may be difficult to trace if they have moved, or they may no longer be alive. It is also debatable whether or not it is always appropriate to seek further informed consent for data to be re-used. An example is where a supplementary analysis is carried out by the same researcher who collected the data in order to follow up an issue from the primary research in more depth: in this case, re-contacting participants may be an extra expense and unnecessary burden on informants as well as a limitation on the scope of the analysis that the primary researcher is allowed to undertake.

An alternative approach to gaining informed consent is to obtain it prospectively, at the time the data is collected. This approach has received more attention as primary researchers' awareness has grown of the possibility that qualitative data sets might be archived or otherwise shared in the future, and of the associated need to inform research participants of this and their legal rights, and also to give them an opportunity to limit access to the material if they wish. Qualidata (undated a, undated b; see also Rolph, 2000) has provided advice for primary researchers on how to obtain permission from interviewees for data archiving. It suggests that researchers approach this in the same ways that they seek participants' permission to take part in the original research, building it into existing procedures for gaining informed consent. Based on this principle, the overall process of gaining consent in a study where the resulting data may be archived, informally shared or personally re-used will involve the following considerations:

- that potential participants are accessed in appropriate ways (with the permission of parents for children aged under sixteen, and other authorities such as ethics committees, schools and employers where appropriate);
- that written information is provided on the aims of the primary study in clear and appropriate language (allowing for age, literacy, visual impairments, learning difficulties and so on);
- that, where possible, appropriate written agreements are obtained from participants stating that they understand and agree to what taking part in

the primary research involves and how the data will be used in this context (checking and affirming this at intervals);
- that, where possible and appropriate, further written agreements are obtained from participants stating that they understand and agree to the data being kept for possible use in other research, together with any conditions they wish to impose on this;
- that participants are provided with opportunities to obtain further information where required;
- and that enough time is allowed for consent to be properly informed and obtained.[11]

Qualidata's guidelines further suggest that participants could be given options concerning data archiving, such as whether or not they would like the data to be anonymized (Qualidata, undated a). They also provide a sample consent form and information on techniques for informing people with learning difficulties about what data archiving involves in practice, such as showing participants examples of how research material can be anonymized and re-used (Rolph, 2000).

Gaining consent for possible secondary research at the time of the original interview effectively expands the process to include not only participation in the immediate primary study but also the use of data in the future. Alderson (1998; see also 2001) identifies three levels of consent associated with this practice: first, consent to participate in the present, primary study; second, consent over the use of the data in the context of the primary research, including associated dissemination of the work; and, third, consent to make the raw data more widely available for possible re-use in future research.[12] An advantage of this approach is that researchers can discuss with participants the option of data archiving, including issues such as where the data would be stored and who would have access to it. Participants can also be given the opportunity to impose conditions limiting access to the data (for example, closing access to the data for set time period). However, this approach is not without problems. In particular, there is the difficulty of anticipating exactly how the data may be re-used in the future, by whom and for what purposes(s). As Alderson observes, can consent be regarded as fully informed when it is so open-ended? A related problem with this approach is that, while primary researchers and archivists can control access to the data, ultimately they cannot guarantee that it will not be re-used inappropriately or insensitively.

In the secondary studies reviewed, it was difficult to establish whether and, if so, how informed consent had been sought because of the lack of published information provided on this aspect of the research. Only two studies reported obtaining informed consent. In the first, informed consent was obtained retrospectively from the research participants (Clarke-Steffen, 1998). By contrast, in the other study, 'provision was made for the possibility of secondary analysis in the consent form' used for the primary research by the second author (Szabo and Strang, 1999: 72). In the remaining studies, it is not clear if informed consent was obtained but not reported or, alternatively, if it was not sought (perhaps because it was presumed that the original consent extended to the secondary research, or because it was not considered necessary).

Confidentiality

In the course of obtaining informed consent researchers and research participants typically enter into some kind of confidentiality agreement over who will have access to the raw data and knowledge of the identity of the informants, and the circumstances in which either of these will be disclosed to third parties. Raw data is normally confidential to all those involved in its collection and processing (that is, members of the primary research team, transcribers and/or data inputters). In quantitative research, data are usually de-identified during processing and presented in aggregate form in reports. Similarly, in qualitative research, textual data are usually redacted to protect informants' identities from being revealed and other data (such as photographs) are disguised or only used with the permission of the subjects concerned. However, in some situations, researchers' duty of confidentiality may conflict with other moral or legal imperatives. They may, for example, feel obliged to report criminal offences, whether or not required to by law; they may also have to break confidentiality agreements and disclose information if so requested by a court.

The principle of confidentiality is generally recognized in professional organizations' codes of practice. In the ethical guidelines of both the American Sociological Association and the Association of Social Anthropologists of the UK and the Commonwealth this topic is specifically addressed in relation to data sharing. The former recommends that:

> Sociologists share data in a form that is consonant with research participants' interests and protect the confidentiality of the information they have been given. They maintain the confidentiality of the data, whether legally required or not; remove personal identifiers before data are shared; and if necessary use other disclosure avoidance techniques. (American Sociological Association, 1997: Section 13.05, (c))

The guidelines further state that, when collecting confidential information, researchers take into account the possible long-term uses of the data and the limits of confidentiality. Similarly, the Association of Social Anthropologists of the UK and the Commonwealth and the British Sociological Association recognize that there are legal limits on the extent to which confidentiality can be protected. As the British Sociological Association warns:

> Research data given in confidence do not enjoy legal privilege, that is they may be liable to subpoena by a court. Research participants may also need to be made aware that it may not be possible to avoid legal threats to the privacy of the data. (British Sociological Association, undated c. 2002: (c))

It therefore recommends that researchers explain the limits of confidentiality to research participants and do not give 'unrealistic' guarantees of confidentiality.

As previously noted, the legal issues relating to confidentiality and copyright in social research were examined by the law firm Allen and Overy (1998) on behalf of the Economic and Social Research Council. It found that researchers in England and Wales may be subject to a general 'duty of confidentiality' which prevents them from disclosing information believed to have been given in confidence, unless they are authorized to do so by the informants. Moreover, where it is explicitly agreed that the information supplied will be kept confidential, this may constitute a legal contract. Based on this legal advice, the Economic and Social Research Council (undated, c. 1998: Section 6) suggest that all data are 'thoroughly anonymized' when placed in an archive 'unless interviewees have consented otherwise'. Likewise, in its guidelines on legal and ethical issues in interviewing, Qualidata recommends that all personal information should be kept confidential, whether or not a pledge of confidentiality has been given, except where individuals want to be identified; it also advises that what is meant by 'confidential' is carefully defined and that written agreements are obtained wherever possible (Qualidata, undated a).

Traditionally, the main strategy for preserving confidentiality is through anonymizing personal information contained in data sets. In the secondary analysis of statistical data, great emphasis is placed on both de-identifying data and preventing linkages across different data sets at the level of the individual.[13] In the secondary analysis of qualitative data, various strategies have been developed for safeguarding confidentiality in accordance with ethical and legal imperatives (particularly the Data Protection Act – of which more below). These include the anonymization of data accomplished by deleting real names or using pseudonyms, and by disguising other identifiers. However, the anonymization of qualitative data does give rise to a number of problems. In particular, the process of anonymizing textual data is time consuming and expensive; it also presents technical difficulties in the case of audio and visual data. Efforts to disguise the identity of informants may also spoil and distort the data through the inappropriate use of pseudonyms and the excessive editing of details and contextual information.[14] Accordingly, other strategies for protecting confidentiality in qualitative research have been employed or proposed. These include: closing access to data sets for a period of time; restricting access to data sets; and imposing conditions on users who access data sets (Corti, 1999; Qualidata, undated a). Qualidata and various archives also emphasize the responsibility of secondary researchers not to breach confidentiality by using any identifiable information in published work (Backhouse, 2002; Qualidata, undated a). The UK Data Archive, Murray Research Center Archive and other depositories require data users to sign agreements to this effect.[15] At the 'German Memory' archive, data is not anonymized but the archivists impose strict conditions on those who can access and re-use the data (Leh, 2000).

Copyright

Social researchers operate within complex legal frameworks which vary from country to country and are subject to change. In general, the relevance of these laws has been unclear and rather glossed over in professional organizations' guidelines for social researchers. Notable exceptions include the Oral History Society which has produced guidelines on copyright (Ward, undated) and the Association of Social Anthropologists of the UK and the Commonwealth (1999). The latter's ethical guidelines include a section on 'Participants' intellectual property rights' which I quote from at length to show the level of advice provided:

> It should be recognized that research participants have contractual and/or legal, interests and rights in data, recordings and publications, although rights will vary according to agreements and legal jurisdiction.
>
> (a) It is the legal obligation of the interviewer to inform the interviewee of their rights under any copyright or data protection laws of the country where research takes place, and the interviewer must indicate beforehand any uses to which the interview is likely to be put (e.g., research, educational use, publication, broadcasting etc).
>
> (b) Under the UK Copyright Act (1988), researchers making audio or video recordings must obtain 'copyright clearance' from interviewees if recordings are to be publicly broadcast or deposited in public archives. Any restrictions on use (e.g., time period) or other conditions (e.g., preservation of anonymity) which the interviewee requires should be recorded in writing. This is best done at the time of the interview, using a standard form. Retrospective clearance is often time-consuming

or impossible where the interviewee is deceased or has moved away. (Association of Social Anthropologists of the UK and the Commonwealth, 1999: Section I, item 7)

Further guidance on the relevance of copyright law in social research has been provided by the Economic and Social Research Council (undated, c. 1998) and Qualidata (undated c) based on legal advice from Allen and Overy (1998). These guidelines explain that, in England and Wales, various kinds of literary works are protected by copyright law as set out in The Copyright, Designs and Patents Act 1988. Under this legislation the copyright of original works (which include recorded interviews) is owned by the author of the words (the interviewee).[16] Where the speaker is an employee of an organization and is talking about their employment and/or is interviewed at work, the employer may also have an interest in the copyright of the work. So too may the researcher who made the recording and the transcriber who produced the transcript. In addition, where a researcher is employed by a University to conduct a study, the copyright may rest with the University, or with the external organization that funded the work. Likewise, reports of the work may be the copyright of the University, as the employer of the researcher, or possibly the funding organization who commissioned the work. Finally, any databases created during the study could be the copyright of the researcher or agency who created them.

Copyright owners have legal rights over how the material is used. In particular, they can prevent others from making unauthorized copies or exploiting their work. They also have moral rights to be identified as the author of the work and that the work is not treated in a derogatory way. Ownership of copyright can be licensed by the owner (by explicit consent or implied authority) or it can be permanently assigned to others (in writing). With regard to the ownership of interview material, the Economic and Social Research Council advises that:

> in order to be able to rely on authority/licence, the interviewee should be given the opportunity to consent to the reproduction of substantial verbatim quotes by a researcher as publication by researchers in the absence of such consent is an infringement of copyright. This applies whether the words go down on paper, audio or videotape. Ideally, the interviewee should be informed at the interview of the broad purposes to which the data will be put. If this is not possible, then permission should be obtained after the interview. (Economic and Social Research Council, undated, c. 1998: Section 3)

Alternatively – and this is what the Economic and Social Research Council recommends – researchers can request that copyright ownership is assigned in writing to the researchers or to an archive, giving the new owner unconditional use of the material.[17]

These guidelines have a number of implications for both secondary and primary researchers. In particular, as the original Allen and Overy (1998) report suggests, The Copyright, Designs and Patents Act 1988 allows for some use of literary works by non-copyright owners under what is termed 'fair dealing'. However, this does not allow for substantial quoting or multiple copying of data and hence the advice is to obtain consent for all uses of data. Another important issue which remains to be addressed is whether and, if so, how groups such as children and people with learning difficulties can be adequately informed about copyright issues and included in decisions regarding the ownership of their original works.

Data protection

Other relevant international and national laws affecting social researchers' use of data include those on data protection and privacy. In the United Kingdom, the Data Protection Act 1998 (implemented in March 2000) governs the use of personal data held on computer and on paper. It gives people rights over the use of personal information and their access to it, adding to those of the Data Protection Act 1984. Accordingly, the Economic and Social Research Council's (undated, c. 1998) guidelines on copyright and confidentiality advise that researchers should inform interviewees about the purpose of collecting the information and who will have access to it, and obtain their agreement to these uses of their personal data. This is consistent with wider advice on the process of obtaining informed consent, although where personal information is anonymized this is exempt from the Data Protection Act 1998 as it is no longer defined as 'personal' data.[18]

Defamation

The Economic and Social Research Council's (undated, c. 1998) guidelines also briefly highlight the potential relevance of laws on defamation. According to the Defamation Acts of 1952 and 1996, one is not allowed to publish defamatory material about a person or organization, unless the statement is true or considered fair comment on a matter of public interest (Allen and Overy, 1998). This law could apply to comments made in an interview about a person or a company which, if published or put in an archive, could lead to the researcher being accused of libel. The Economic and Social Research Council therefore recommends that the subjects of any potentially defamatory statements are not identifiable from the data.

Conclusion

The extension of secondary analysis from quantitative to qualitative data has been accompanied by growing concerns over the ethical and legal aspects of retaining and re-using the latter data in further research studies. This has been reflected in the development of guidelines on these issues and in the nature of the revisions made to the codes of practice of some learned societies. However, these issues were seldom discussed in the reports of actual qualitative studies and hence it is not clear how they were addressed in practice, if at all.

Further research is required to examine how researchers can best obtain informed consent, maintain confidentiality, and negotiate copyright agreements concerning the collection and use of data in primary and secondary qualitative studies. There is also a need for related research exploring research participants' views on the circumstances in which qualitative data may and may not be re-used, by whom, in what form, for what purposes, and with whose permissions.[19] Only then will it be clearer whether any harm can be done by re-using qualitative data and, if so, how this can be avoided. Moreover, as Hood-Williams and Harrison (1998) suggest, many of the issues raised over the re-use of qualitative data in secondary research apply equally to the treatment of data in primary research, particularly in open-ended research designs where the precise focus of the analysis emerges in the course of the work, after ethical approval has been granted and informed consent has been obtained. The aforementioned research agenda could therefore be usefully expanded to explore research participants' views on the conduct of qualitative and quantitative research in general and issues such as: the ways in which informed consent is best obtained in different types of qualitative and quantitative studies and from particular groups of research participants; what confidentiality means in practice; the ways in which different types of qualitative and quantitative data are stored and shared in accordance with access agreements; the perceived benefits, costs and risks of taking part in primary social research projects and making data available to be used in further research studies; and to how participants could be informed about and involved in other studies using the data they provided.[20]

Notes

1. Unfortunately, it was not possible to examine guidelines which were not available in English, such as those developed by archives in Scandinavia and several other European countries.
2. These are re-printed in Connor and Fuenzalida-Puelma (1990).

3. All references are to the *current* electronic versions of these organizations' codes of practice available at the relevant URL on the specified date of access. Several of these organizations have recently reviewed their guidelines, or are in the process of reviewing them, and hence they are subject to change.

4. The National Children's Bureau's guidelines are based on and extend those of the British Sociological Association (National Children's Bureau, undated).

5. At the time of writing, the role of LRECs and MRECs was under examination as part of the Department of Health's review of the governance of health and social care research. A major focus of the review is on the process of obtaining informed consent (following the revelations about clinical research practices at Bristol and Alder Hey hospitals involving the retention of medical specimens and body parts from deceased patients for possible research without the full authorization of patients or their relatives) and it is likely to have implications for social as well as clinical research practices.

6. The Office for Protection from Research Risks (OPRR) has produced a guidebook for the reference of IRB members, researchers and administrators which deals with ethical issues and regulations concerning research involving human subjects in the United States (Office for Protection from Research Risks, undated).

7. Research carried out without participants' knowledge and consent is often referred to as 'covert' research. Reynolds (1982: 79) has suggested that archival research is a form of covert research where 'neither the individuals originally providing the data, nor anybody else, was aware that the data would be used in the investigator's specific research project'.

8. See The Royal Liverpool Children's Inquiry Report (also known as the 'Alder Hey Report' and 'The Redfern Report') 30 January 2001 (Michael Redfern QC) and the Bristol Royal Infirmary Interim Report ('The Kennedy Report') 2001 (Sir Ian Kennedy).

9. However, there is an interesting clause in the American Sociological Association's ethical guidelines relating to 'Unanticipated Research Opportunities' which is relevant to researchers who, for example, decide to follow up their primary research with a supplementary analysis of the same data set:

> If during the course of teaching, practice, service, or non-professional activities, sociologists determine that they wish to undertake research that was not previously anticipated, they make known their intentions and take steps to ensure that the research can be undertaken consonant with ethical principles, especially those relating to confidentiality and informed consent. Under such circumstances, sociologists seek the approval of institutional review boards or, in the absence of such review processes, another authoritative body with expertise on the ethics of research. (American Sociological Association, 1997: Section 13.02)

10. An example of a consent form used for this purpose is reproduced in Hinds et al. (1997: appendix B).

11. This should include time at the beginning, during and at the end of an interview to obtain the necessary consent and permissions, and to confirm these when appropriate. In addition, some aspects of the research, such as options concerning the use of data in the primary study and data archiving, are arguably best discussed at the end of an interview when the informant is fully aware of what she or he has contributed. For a discussion of the process model of gaining informed consent in primary qualitative research, see Ramcharan and Cutcliffe (2001).

12. It is not clear whether Alderson is referring to raw data which has not been modified in any way, or to anonymized versions of these data sets which have been modified for the purposes of data sharing and secondary analysis.

13. For example, the Inter-university Consortium for Political and Social Research (2002) guidelines state that data should be anonymized before archiving and any personal identifiers removed or amended if necessary.

14. For a useful discussion of the use of pseudonyms in social research see Barnes (1979) Chapter 7: 'Communicating results'.

15. The UK Data Archive's 'Access agreement for individuals' (2001, version 3) is available at www.data-archive.ac.uk/orderingData/agreements/accessIndividual.pdf [accessed 18/02/2003]. A sample copy of the Murray Research Center archive's 'Memorandum of Agreement' between the archive and data donors is reproduced in James and Sørensen (2000: appendix).

16. This right expires 70 years after the death of the author.

17. An example of a 'Clearance and copyright form' is provided in the relevant guidelines of the Oral History Society (Ward, undated). Qualidata also provide an example of a combined consent and copyright transfer form for adaptation to the requirements of particular studies. This is available at http://www.qualidata.essex.ac.uk/about/document.asp [accessed 18/02/2003].

18. Additional guidelines for records managers and archivists on the Data Protection Act have been produced by the Public Records Office (2000).

19. For example, researchers at Qualidata have observed that some informants who provide life histories want to be identified with the material and that it is important to give them the option of being identified (Thompson, 2000a) and that they are 'seeing more instances of informants being more than willing for their data to be archived even without preservation of confidentiality, contrary to the expectations of many researchers, if [informants are] given the opportunity to approve transcripts' (Backhouse, 2002: 33).

20. The issue of whether patients wish to be consulted about researchers' access to their medical records has been explored by Willison et al. (2003).

6
Modi Operandi

Design	90
Selection of data set(s)	91
Analysis	96
Quality assurance	99
Reportage	102
Reflexivity	104
Conclusion	106
Notes	107

While the epistemological and ethical aspects of qualitative secondary analysis have received increasing attention in the literature, there has been relatively little work on the practicalities of re-using qualitative data. Consequently researchers who have carried out secondary studies have not had a great deal of information on the 'doing' of qualitative secondary analysis to guide them. In this chapter, I draw on reports of the methods used in the secondary studies reviewed, together with related commentaries by some of the authors on their experiences of re-using qualitative data, in order to examine what the process of doing secondary analysis involves in practice. Various aspects of the research process are explored, ranging from the decision to do social research using the methodology through to the reporting of the work undertaken. I consider how approaches to re-using qualitative data vary depending on the source of the data and other factors, and how the methods used in qualitative secondary analysis compare to those used in primary qualitative research. The aim is to critically describe and review existing practice rather than to produce a guide on 'how to do' qualitative secondary analysis, although it is hoped that this work will help to inform the future development and use of the methodology.

Design

In theory, there are various reasons why researchers might decide to carry out research using secondary rather than primary research methods. For example, it may be that, for logistical or other reasons, relevant data cannot be collected on a primary basis and so available data have to be used where this is feasible. This applies to some kinds of historical research and to hard to reach populations, where researchers may only have access to pre-existing data; it also applies to students for whom the opportunities and resources to do primary research may be limited. Of course, secondary research methods may also be preferred where relevant, good quality data sets already exist, providing researchers with a ready-made source of material and negating the need to collect more data.

While the circumstances and rationale for doing secondary analysis were not usually fully reported in the studies reviewed, it was apparent that many of them had been designed in order to follow up new or related questions arising from the researchers' previous work (as in the examples of supra and supplementary analyses).[1] Thus, often the questions explored appeared to have been endogenously derived from the researchers' previous, primary, analysis of a data set. The same data set was then re-used in the secondary study (redefined as a secondary data set) in order to explore aspects of the data which were not examined in the initial investigation. Sometimes only this data set was re-used, in full or in part, with no other data being involved; in other cases, two or more pre-existing data sets were re-used, or additional primary or secondary (naturalistic) data were collected to add to the researchers' own data (as in amplified and assorted analysis, see Chapter 3). It appears, therefore, that these secondary studies were developed as by-products of the researchers' primary work, and that secondary analysis was employed as an adjunct to the primary research, rather than as an alternative to it.

In the few cases where analysts were re-using other researchers' data and had not been involved in the primary work, it was not clear from the reports whether the data sets had been found and used to address preconceived, exogenous, research questions or, alternatively, whether the ideas for the studies had emerged from the researchers' knowledge of the material. For instance, it was not reported whether a particular data set had been selected following a trawl of data sets to see if any were available which fitted a pre-defined research question, or whether the question arose from the researchers' initial perusal of the data set concerned, or from other indirect knowledge of the data set. With the increase in the number of qualitative data sets being archived, it is now possible that researchers can use these resources in these various ways (indeed, the

Economic and Social Research Council in the United Kingdom currently requires applicants for research funding to check and affirm that their proposed primary research cannot be undertaken using existing secondary data). However, unlike quantitative researchers, qualitative researchers do not yet have the option of developing studies based around omnibus qualitative data which have been specially collected to facilitate secondary research (with the exception of historical collections). Thompson (2000a: para 47) has argued that more 'basic' qualitative data sets need to be collected 'for the general use of qualitative researchers'.

Selection of data set(s)

The process of selecting a data set for secondary analysis can be divided into two stages. First, a search for relevant data sets needs to be conducted if one has not already been found by the researcher. Second, potential data sets then need to be checked to see if they are re-usable for the purposes of the proposed study. How this was approached in the studies reviewed and the options available for researchers are further examined below.

Finding relevant data sets

Although relatively few of the studies reviewed involved the re-use of other researchers' data, an increasing number of data sets from various areas of social research have been deposited in archives in recent years and are now available for possible secondary analysis. Researchers can access information on the availability and content of qualitative data sets deposited at archives across the United Kingdom through an on-line catalogue (Qualicat) set up and managed by Qualidata.[2] Qualicat provides useful information on deposits, such as the names of the investigators, main topics of the investigation, the type and quantity of research material, format of the material, location of the deposit, and any conditions concerning the use of the data. Many archives in the United Kingdom and other countries also provide on-line as well as in-house information on their own deposits. Some archives have searchable electronic catalogues, such as the 'German Memory' archive which has a collection of over 1,500 life history interviews and every interview is described on up to 127 criteria in an electronic 'reference book' (Leh, 2000).

Even where good information is provided on the content of archived data sets, the data itself will have to be scrutinized in order to establish that it is re-usable for the proposed study. This is likely to require a visit to the archive as it is rare that data sets are available for consultation on-line. If more samples of archived data sets were available for inspection on-line this would help researchers to both identify potential data sets and to provisionally assess whether or not the material was fit for the purposes of the secondary study. Of course, the digitization of whole data sets could also allow the material to be accessed more easily. As Fielding (2000) has pointed out, this is not an 'impossible dream' as the technology for it exists at present. Indeed, the Centre for Social Anthropology and Computing (CSAC) at the University of Kent at Canterbury has already published anthropological fieldnotes and data on the Internet. For example, part of Stirling's longitudinal fieldwork in two Turkish villages conducted 1949–94 is available on-line[3] and the remainder is available to 'qualified scholars upon application' (Zeitlyn, 2000; para 7). Selected qualitative data have also been published on the Internet as part of the Oxford-based database of patients' experiences (DIPEx) project, but not the whole data sets as such.[4]

Assessing the re-usability of data sets

Apart from commenting on the suitability of the data sets that they had used, particularly in terms of the 'fit' of the data with the research question, researchers seldom described exactly how they had established that the data were re-usable for the purposes of the proposed study. Thus, it was not clear what, if any, checks they had made to establish that the data sets were re-usable.[5] However, as part of their work exploring the potential of qualitative secondary analysis, Hinds et al. (1997) developed an assessment tool for checking the accessibility, quality, completeness and fit of data sets (Hinds et al., 1997: appendix A). The tool was designed to assess the re-usability of data sets collected by other researchers and for this reason is less suitable for those who may be interested in re-using their own data. I have therefore modified the tool and used it to form the basis of the following guidelines for checking the accessibility, quality and suitability of various sources of data (see Table 6.1).

Accessibility

Some data sets are easier to access than others, depending on who owns and/or controls access, where the data are held, and in what format. Where a sole researcher is planning to re-use his or her own data, access is not likely to be a concern. However, where two or more researchers are combining or pooling their respective data sets, as in amplified analysis, the data will have to be physically and/or electronically merged to facilitate the secondary analysis; the space required by, and the compatibility

Table 6.1   Guidelines for assessing the re-usability of qualitative data sets

Accessibility

Where, when and how can the data set be accessed?
Are all the data accessible, or only part of the data set (e.g. transcripts but not tapes)?
Have informants given informed consent for the data to be used for the purposes of the proposed study?
Are there conditions, or terms of usage, associated with the use of the data set?
Can the primary investigator(s) be consulted, if desired?

Quality

Is the data set complete for the purposes of the secondary study (i.e. no or minimal missing data)?
Has the data been recorded fully and accurately (e.g. accuracy of transcriptions)?
Have any data been modified (e.g. to preserve anonymity) and, if so, how?
Has the data set been adequately prepared for possible secondary analysis?
Is the meta-documentation of the data set sufficient for the purposes of the secondary analysis?
Was the primary study well designed and executed?

Suitability

Is the data set 'fit' for the purposes of the proposed research?
Is the sample adequate for the proposed research?
Are there sufficient data to address the proposed question?
Is the type, and format, of the qualitative data compatible with the proposed research?
Can the data be combined or compared with other data sets, if required?
Is the age of the data set appropriate?

Source: adapted from Hinds et al. (1997: 420–21) with permission of Sage Publications.

of, electronic data will also need to be checked to see if it can be shared and re-used in this format. Physical access is also an issue for researchers proposing to use data sets in archives, especially where they have no option but to consult the material *in situ*. This differs from the secondary analysis of quantitative data where such material is generally shared electronically. However, more qualitative data sets are now being scanned and archived in digital form which should facilitate electronic access in the future.[6]

In many cases, researchers will need to consult with relevant parties in order to gain access to the data sets. For example, those who plan to re-use their own data may want to inform any colleagues who were involved with the primary project in case they object to, or also want to be involved in, the secondary study. Similarly, in informal data sharing, the primary and secondary researchers may wish to have a written agreement setting out the terms of access and re-use. For example, Szabo and Strang had a letter which:

> outlined the rights, responsibilities, and obligations of the original and secondary researchers. Included in the letter was a description of the data that were accessed (eg, interviews, demographic data), method of access (ie, via computer software), and provisions for reference citations in publications and presentations. (1999: 72)

Researchers planning to use an archived data set may have to apply for permission from the archive, which in turn may seek the approval of the principal investigator who deposited the data. Access may also be conditional on the proposed study having ethical approval from a recognized body. In addition, data donors and/or participants may have imposed restrictions on the usage of the data, such as only allowing access after a specified time-span or only consenting to the data being used for particular research purposes. Confidentiality agreements may also limit researchers' access to transcripts and not tapes of interviews because of difficulties anonymizing the latter data; they may also prevent secondary researchers from re-contacting informants. Finally, as well as gaining access to the actual data, researchers may need to be able to consult with the principal investigator or team who conducted the primary research. This is particularly important when further knowledge of the context in which the work was undertaken is required, or when there are gaps in the meta-data of the study.

Thus, the process of gaining access to data sets and data donors can be complicated and time consuming. Although there has been no empirical research on the relative costs of primary and secondary research using quantitative or qualitative data, it has been generally assumed that it is cheaper to share quantitative data than to collect the data using primary methods (Fienberg et al., 1985). This is also likely to apply to qualitative data, although the difference may be less marked and will vary depending on the respective involvement of primary researchers, secondary researchers and/or archives in preparing the data for archiving. It is also likely that qualitative secondary analysis using archived data sets is more expensive than using personal sources of pre-existing qualitative data, although the quality of the former data (as a secondary resource) may be higher given that it has been specially prepared for re-use.

Quality

In qualitative as well as quantitative secondary analysis, the completeness of the data is an important criterion for assessing the quality of a data set. Gauging the extent of missing data is, however, harder in qualitative research given the relatively unstructured nature of the data. For example, in studies where semi-structured interviews are used, the same questions may not have been posed to all informants, and the questions may have been asked in a different order and pursued to varying degrees of depth. This makes it difficult to check the extent to which particular topics of interest are complete. A related quality criterion is whether all the data are included or not. For instance, it is possible that only transcripts and not audio-tapes of interviews are available because of the difficulties of anonymizing the latter data. This will not only prevent the accuracy of transcripts from being checked but may also affect the interpretation of the material, particularly where this depends on audible features such as intonation and emphasis. The interpretation of data may also be hampered where tapes and/or transcripts have been extensively edited to remove or disguise real names, places, and other identifying material.

From a secondary data user's perspective, the overall quality of a data set will further depend on how well it has been preserved and the process of the primary research was documented. Increasingly, primary studies are planned and executed in anticipation of the resulting data set being archived for possible secondary analysis.[7] Indeed, in an innovative study in Canada, an archivist was employed as part of the primary study team to prepare the data for archiving in the course of the primary work (Humphrey et al., 2000). However, in the past, data sets have been archived as an afterthought and hence the re-usability of the material will depend partly on how well the data collection and processing was routinely managed in the course of the primary work and partly on how well this process was retrospectively documented for archival purposes.

Finally, the quality of the design and conduct of the primary work itself is another important consideration. Hinds et al. (1997) suggest that, in assessing the re-usability of other researchers' data sets, the credentials and training of those who undertook the work, as well as the conduct of the interviews and how the sample was selected, are examined. They therefore imply that qualitative secondary analysis is best carried out using data from studies where the skills of the researchers can be established and the methods employed are found to be satisfactory.

Suitability
The third and final set of criteria relate to whether or not the data are suitable for the proposed study. As I have observed, the fit of the research question with the data is less likely to be a problem where the aim of the secondary study is to investigate issues arising from a previous analysis of the same data set. It is, however, likely to be a major consideration where

researchers are searching for possible data sets to use to investigate a preconceived research question. Two related concerns are whether the size and composition of the sample are adequate for the purposes of the secondary study, and whether there are sufficient data for the proposed study. Even where the idea for a secondary study emerged from a prior analysis, this does not necessarily mean that the data set contains enough material on which to develop a secondary study, particularly where only a sub-sample of the material is germane. If the data are insufficient, researchers can consider adding more data from other sources, or undertaking primary research as well, as in the examples of assorted analysis described in Chapter 3. Likewise, the type of qualitative data may be important depending on the nature of the proposed secondary study. Some secondary research questions may, for example, be best explored using interview rather than observational data, or using open-ended rather than more directed interviews.[8] Finally, the age of the data set(s) is another important consideration in determining the suitability of the material for use in a secondary study (Hinds et al., 1997).

Overall, the process of establishing the re-usability of data sets will vary depending on whose data is being re-used. An advantage of informal data sharing (as well as re-using one's own data) is that one has information directly from the data donor (or from one's own experience) about the design and conduct of the primary study. This information can help one to determine more easily, and with confidence and speed, the quality and suitability of the data set(s) being considered. For an independent researcher using archived data sets, these matters are more difficult to consider, as they require preliminary assessment of the data and metadata, and possibly consultation with the depositors, in order to establish whether or not they are good for the proposed secondary study. Finally, depending on which source of data is being used, researchers may also need to agree terms of access and usage of the data in contracts with either the primary researchers in the case of informal data sharing (Hinds et al., 1997) or with the data archive (see Leh, 2000).

Analysis

There are various theoretical approaches to, and methods of, qualitative data analysis, some of which were developed in order to study non-naturalistic data collected in the course of primary research and some of which were developed for the study of naturalistic material.[9] In the studies reviewed, an array of analytical approaches were used, including 'grounded theory' (Clarke-Steffen, 1998; Hutchinson, 1987, 1990; Jenny and Logan, 1996;

Jones, 1997; Kearney et al., 1994a, 1994b; Knafl et al., 1995; Logan and Jenny, 1990; Messias et al., 1997; Robinson, 1993; Sandelowski and Black, 1994; Szabo and Strang, 1999; Thorne, 1990c; Thorne et al., 2000), varieties of 'content analysis' (Breckenridge, 1997; Ferrell et al., 1996; Pickens, 1999; Princeton, 1993; Thorne, 1988) and 'narrative analysis' (Gregory and Longman, 1992; Hall et al., 1992; Knafl et al., 1995). The approaches that were adopted often mirrored those that the researchers had initially used in the primary work to gather the data. In general, few problems were reported with the methods used to analyse the data, or with the amenability of the data to secondary analysis. However, as I show below, some exponents of grounded theory did discuss how they had adapted the techniques associated with this approach for use in the context of secondary research. The secondary analysis of qualitative data also involved some practices which differ from those used in primary qualitative data analysis.

Adaptations to grounded theory

In primary research using grounded theory, decisions regarding what data to collect are revised as the study progresses, based on emerging theories derived from the coding and analysis of the data already collected (Strauss and Corbin, 1990). This process of 'theoretical sampling' is potentially problematic in secondary analysis because the data are already 'given'. However, as Strauss and Corbin have indicated, theoretical sampling can be accomplished in other ways. They advise that, when re-using archival data, 'You sample exactly as you do with interview or observational data, and then there is the usual interplay of coding and sampling' (1990: 188). Similarly, they suggest that when re-using the 'collected interviews or fieldnotes' of other researchers, 'A grounded theorist can code these materials too, employing theoretical sampling in conjunction with the usual coding procedures' (1990: 189).

This was accomplished in two ways by some of the researchers who had used grounded theory in the secondary studies reviewed. One approach involved selecting data from within the available sample. This strategy was used by Kearney et al., in order to 'seek out examples of different experiences and contexts as the analysis progressed' (1994b: 145; see also Kearney et al., 1994a). Szabo and Strang (1997: 68) also employed it to select 'the type of informant required in order to saturate the categories that emerged from the data'. While this approach allows analysts to work flexibly within the data sets, as the authors observe, a limitation of this *post hoc* approach was:

> our ability to return to respondents to validate our conclusions and/or ask new questions. Therefore, some areas of the analysis are noted to

be less well saturated than others and are presented as such in the findings. (Kearney et al., 1994b: 146)

In the other approach, additional primary data were collected in order to augment the secondary data set. It was used by Hutchinson in her two secondary studies on nurses' self-care strategies (1987) and nurses' rule-bending behaviours (1990), in which a sub-set of the data from her primary research on unprofessional behaviour among hospital-based nurses was re-used together with additional data specifically collected for each study.

Sorting

More generally, in several of the secondary studies reviewed, researchers selectively drew upon the data set(s) which formed the basis of the secondary study. As noted in Chapter 4, sorting is a technique which secondary researchers have often used to manipulate and shape the data set(s) so that it fits with the purpose of their analysis. Thus, sorting was used as a way of separating out qualitative from quantitative data (see Clayton et al., 1999; Hall et al., 1992), and interview from observational data (West and Oldfather, 1995), in order to focus on just one type of data. It was also used to select sub-samples of the study population (see Fendrich et al., 1996; Kearney et al., 1994a; McLaughlin and Ritchie, 1994) and to focus the analysis on particular topics or themes (see Ferrell et al., 1996; Gallo and Knafl, 1998; Gregory and Longman, 1992; Jairath, 1999; Jenny and Logan, 1996). For example, two studies focused on transcripts from larger data sets which were found to contain metaphors (Jairath, 1999; Jenny and Logan, 1996).

Re-coding

Where qualitative data have been previously coded the codes may be removed, revised or retained depending on their relevance and accuracy. In some types of secondary analysis, the raw data is re-coded from scratch. Thus, in pooled case comparison where two or more data sets are merged, previous categorizations and interpretations of the data are set aside so that the 'new analysis begins with a clean slate' (West and Oldfather, 1995: 454). This approach therefore involves some undoing of the original primary analysis in order to shift the focus back to the raw data. More generally, it is not clear from the reports of the secondary studies reviewed whether or not the cleaning of data is standard practice in other forms of secondary analysis. This may well depend on the respective foci of the primary and secondary studies. In supra analysis, for example, the themes of interest are very different from those of the primary study and so new indexes for coding will have to be devised and applied. The same applies to amplified analysis where data are pooled from different studies which were originally coded separately. By contrast, in a supplementary study, a broad theme may have been previously identified and coded in the primary analysis but requires more elaboration and/or linking to other codes for the purposes of the secondary analysis. Whatever the focus of the secondary study, it is likely that certain codes – especially those concerning the socio-demographic characteristics of respondents – are still relevant and could therefore be usefully retained for the secondary study.

Computer-assisted qualitative data analysis

One other way in which the analysis of primary and secondary data differs is that the options for computer-assisted qualitative data analysis are at present more limited in the latter case. This is because older data sets in particular are not always available in electronic form. Accordingly, researchers may have to undertake analysis using manual methods or have copies of the data typed up for input onto computer. Some of the studies reviewed reported using software (including The Ethnograph, NUD•IST and WinMAX) to assist with the analysis (see Angst and Deatrick, 1996; Clayton et al., 1999; Fielding and Fielding, 2000; Gallo and Knafl, 1998; Heaton, 2001; Sandelowski, 1994b; Szabo and Strang, 1999). All of these studies involved the re-use of the researchers' own data except for Fielding and Fielding (2000). Weaver and Atkinson (1994) also used Roth's locally archived data to illustrate their book on microcomputing and qualitative data analysis.

Quality assurance

There is considerable debate in social research over how the quality of qualitative work can be demonstrated and discerned (see Creswell, 1998; Denzin and Lincoln, 1994; Hammersley, 1990; Murphy et al., 1998; Patton, 2002; Pope and Mays, 2000; Reid and Gough, 2000; Smith, 1984; Temple, 1998). For present purposes, three broad perspectives will be distinguished.

According to the first, the same or similar criteria used in quantitative research can and should be applied to qualitative research (Kirk and Miller, 1986; LeCompe and Goetz, 1982). This view relies on the so-called 'foundational' assumption that the actual correspondence of truth claims about the social and natural world can be determined through the application of rigorous scientific methods (Lincoln, 1995; Smith, 1984). In contrast, the second perspective disavows the idea that there is an independent reality which can be distinguished and maintains that alternative criteria are required which better reflect the interpretive nature of truth claims, including those of qualitative researchers. Of the various alternative criteria which have been proposed (see Beck, 1993; Guba, 1981; Lincoln and Guba, 1985; Reid and Gough, 2000), Lincoln and Guba's (1985) criteria of 'trustworthiness', outlined as part of their conceptualization of 'naturalistic inquiry', have been particularly influential. These criteria replaced the established, positivistic, concepts of 'validity', 'reliability', 'generalizability' and 'objectivity' with those of 'credibility', 'dependability', 'transferability' and 'confirmability' respectively. Related procedures have also been developed to enable researchers to establish the trustworthiness of their claims, such as 'thick description', 'member validation', and 'triangulation' (Lincoln and Guba, 1985). However, according to the third and final perspective, these are, in effect, alternative foundational criteria for discriminating the veracity of claims about the social world and as such are incompatible with the purported antifoundational assumptions of naturalistic inquiry in particular, and qualitative research in general (Smith, 1984). This perspective has, in turn, informed the subsequent revision and expansion of the aforementioned criteria, including the introduction of 'authenticity' criteria for use in interpretive research (Guba and Lincoln, 1989; Lincoln, 1995).

Of these three perspectives, the second has been most influential in the development of qualitative secondary analysis to date. For example, writing about the potential of secondary analysis in qualitative research, Thorne (1994: 275) suggests that the procedures commensurate with the second perspective are 'with a few relatively minor exceptions... accessible to the secondary analyst' and are adaptable for the purposes of secondary analysis. Similarly, the review showed that three of the techniques associated with the second perspective, namely triangulation, member validation, and audit trails, have been used in practice to date. The use of these techniques in the studies reviewed is described below, together with examples of how they were sometimes modified for use in the context of secondary research.

Triangulation

In navigational terms, 'triangulation' is a means of locating a position by taking bearings from known landmarks. Triangulation was originally adopted in the social sciences by Campbell and Fiske (1959) in the form of 'multiple operationalism' – the use of multiple independent measurements as a means of minimizing any bias or partiality from the use of a single method and so establishing the validity of knowledge claims (Murphy et al., 1998). The concept of triangulation was subsequently applied and extended to qualitative research by Denzin (1970) who identified four basic types. The first and best known of these is 'methodological triangulation' which involves the use of different methods to examine the same phenomenon. In the second type, 'data triangulation', different data on the same phenomena are collected. 'Investigator triangulation' refers to the involvement of different researchers in the same study. Finally, 'theoretical triangulation' is the use of different theoretical models in the same study. Other types of triangulation have since been identified, including 'analysis triangulation' (Burns and Grove, 1993), 'interdisciplinary triangulation' (Janesick, 1994) and 'data type triangulation' (Miles and Huberman, 1994: 267).

Although the term 'triangulation' was not used to describe the methods used in the studies reviewed, nonetheless similar methods appear to have been employed in some of the studies. For example, different sources of data – including autobiographies, additional primary or secondary research data, and documentary sources – were used in some studies for the purposes of corroborating analyses in a form of data triangulation (see Cohen, 1995; Hutchinson, 1987, 1990; Kidd et al., 1996; Tishelman and Sachs, 1998; Thorne et al., 1999). In other studies, a form of investigator triangulation was used whereby the secondary researchers consulted either the primary researchers who collected the data, or other colleagues, over their interpretation of the data (see Clarke-Steffen, 1998; Szabo and Strang, 1997, 1999). Finally, in one study, the same data were re-analysed using different methods in a form of methodological triangulation designed to verify one of the author's own previous analysis of the data set (Popkess-Vawter et al., 1998).

Member checks

This approach to validating qualitative research findings involves exploring the correspondence between researchers' and members' understandings of a phenomenon. 'Members' include research participants and other representatives of the study population. Lincoln and Guba (1985: 314) proposed that this was 'the most crucial technique for establishing credibility'. Different approaches to member validation have been suggested. Bloor (1997) describes three: the prediction of members' descriptions; the ability of researchers to 'pass' as a member; and the checking of researchers' analyses by members. It is this third form of member validation which was employed in three of the secondary studies reviewed.

Logan and Jenny (1990: 38) 'shared the analysis with the study informants [nurses]' in order to enhance the 'rigor' of their secondary analysis using data from their previous primary work. By contrast, in the other two studies, it appears that other members were substituted for the original study informants. Thus, Szabo and Strang (1999: 72) report that 'a former family caregiver of a relative with dementia was consulted' in their secondary study on this group of carers. Similarly, Thorne et al. (1999: 44) presented preliminary findings of their secondary study to a community advisory panel which included 'consumer advocates from each of the communities involved'. By 'cross-referencing' between the consumer panel and the data set (comprising focus group material and letters from individual women diagnosed with cancer) the researchers were able to 'refine and elaborate on the original themes' (Thorne et al., 1999: 44).[10]

Audit trails

As we saw in Chapter 4, the audit trail has been suggested as a way of enabling the rigour and trustworthiness of naturalistic inquiry to be checked by independent auditors (Halpern, 1983; Guba, 1981; Lincoln and Guba, 1985). Although the audit trail was originally developed for use in the context of primary research, it was used in three secondary studies as a means of assuring the quality of these projects (Logan and Jenny, 1990; Szabo and Strang, 1999; Thorne et al., 1999). However, it is not clear from the reports of these studies whether the audit trails actually facilitated the internal and/or external audit of these studies, nor what the process of keeping an audit trail involved in the context of secondary research. Accordingly, it remains to be established whether or how the sort of audit system proposed by Halpern and championed by Lincoln and Guba is applicable to the process of carrying out qualitative secondary studies.[11]

Reportage

Previous work has highlighted potential ethical and representational difficulties arising from the relatively distant relationship of the secondary researcher to the data. For example, Thorne has observed that:

> Where a primary researcher typically has access to key participants and other sources of consensual validation for both the accuracy and the responsible interpretation of the findings, a secondary researcher will necessarily be somewhat removed from the context of the original research and may be at increased risk for misrepresentation or misappropriation. (1998: 553)

While this issue was not raised in the studies reviewed or in related commentaries, other problems have been identified with the reporting of qualitative secondary studies, particularly with the definition of the methodology employed (Heaton, 1998; Thorne, 1994) and with the accounts of how the research was carried out (Heaton, 1998). These problems were confirmed by this updated review.

'Secondary analysis' was reported to be the methodology used in the majority of the studies reviewed. However, as we saw in Chapter 3, in some cases other terms were used and occasionally the methodology was not defined at all. Researchers who did define their work as secondary analysis did not usually claim to be using a particular type of qualitative secondary analysis, although a few relatively recent studies (see Fielding and Fielding, 2000; Messias et al., 1997; Yamashita and Forsyth, 1998) were described as exemplifying one of the types of qualitative secondary analysis outlined by Thorne (1994) and others (Estabrooks et al., 1994; Heaton, 1998). The variation in the ways in which the studies have been defined may well reflect the fact that qualitative secondary analysis is a relatively new and emerging methodology, as well as the blurry nature of the boundaries between primary, secondary and meta-analysis (see Chapter 1).[12]

Most of the secondary studies provided only brief and often tantalizing details of how the research was carried out.[13] As Thorne (1994) has observed, in reporting qualitative work authors are often limited in the amount of detail they can provide because of space constraints imposed by many journals. This is a particular problem for secondary analysts who may require more space in order to fully describe the background to the study and the respective methods used in the primary research as well as the secondary study.[14] Certainly the under-reporting of the methods used in the studies reviewed severely limited the extent to which the quality of the work could be appraised (although it should be noted that a lack of information is not necessarily indicative of a lack of quality). There is clearly a need for more information to be reported on the conduct of secondary studies, and for more space to be provided by journals for this. While the kinds of information and level of detail required will vary depending on the complexity of the study, some of the basic methodological matters which may need to be reported are listed in Table 6.2.[15]

Increasingly in social research, published articles are being subject to peer review and systematic review using criteria which have been developed to help reviewers appraise the quality of the work and synthesize the findings. Such criteria were originally developed for quantitative research but there is growing interest in developing equivalent criteria for qualitative research and initial guidelines have been proposed (see British

Table 6.2 Examples of possible methodological matters to be described in reports of qualitative secondary studies

The data set(s)

What source of data was used – personal, informal or formal – and why?
What was the aim of the primary research?
What type(s) of data does the data set(s) contain?
How was the re-usability of the data set(s) assessed?

The secondary analysis

What was the function of the secondary analysis (i.e. to investigate a new research question or to verify a previous study)?
(How) did the aim of the secondary study differ from that of the primary research?
(How) was ethical approval obtained for the secondary research?
(How) were the data sorted or pooled for the secondary analysis?
(How) were the data re-coded for the secondary analysis?
What analytical approach was used and why?
(How) was the quality of the secondary analysis assured?
(How) was the primary researcher(s) involved in the secondary study?
(How) did the secondary researcher's previous (lack of) involvement with the primary research from which the data were derived affect the analysis?
What were the main strengths and limitations of the secondary analysis?

Sociological Association Medical Sociology Group, 1996; Elliot et al., 1994; Popay et al., 1998). Although these guidelines have been designed to facilitate the review of primary research, they also provide further help for authors of secondary studies on what information they need to include in reports in order to demonstrate the quality of their work (see Appendix B for the guidelines on peer review produced by Elliot et al., and the British Sociological Association Medical Sociology Group).[16]

Reflexivity

Reflexivity in primary qualitative research generally involves the self-examination of how research findings were produced and, particularly, the role of the researcher(s) in their construction (Davies, 1999; May, 2002; Murphy, et al., 1998; Seale, 1999). It can take different forms, ranging from the 'confessional accounts' of researchers' experiences in the field through to post-modern analyses which recognize and celebrate the manifold possible interpretations of the material and so downplay the 'authority' of the analyst(s) who authored reports of the research (Seale, 1999). Surprisingly, given the complex circumstances in which secondary studies are produced, reflexivity was not a particularly strong feature of the studies reviewed. Apart from occasionally discussing the limitations of working with 'given' data, authors rarely discussed issues such as the genesis of the project, the issues involved in re-using one's own data or other researchers' data, how analysing data compares in the context of primary and secondary work, the implications of involving or not involving the original researchers and/or research participants in the original study, and the different contexts in which the data were collected and interpreted first for primary and then for secondary purposes.[17] A notable exception was Atkinson (1992) who re-used his own data in order to demonstrate different ways of reading data. This involved juxtaposing his original analysis of the data in its categorized and fragmented form, with an alternative narrative analysis of the data as it appeared in his notebooks.

In general, it was therefore hard to discern from these reports to what extent researchers had been reflexive in the course of developing their analyses. However, a few authors have also published separate commentaries on their experiences of doing qualitative secondary analysis which better illustrate the ways in which reflexivity has been exercised (see Hinds et al., 1997; Mauthner et al., 1998; Szabo and Strang, 1997,1999). Thus, Hinds et al. (1997) describe the conduct of two secondary analyses, including one study which was abandoned. Their account includes a focus on the problems posed by the respective relationship of the primary and secondary analysts to the data. In one of the studies, the secondary analysis was discontinued after failure to establish satisfactory levels of inter-rater reliability in the secondary coding of the transcripts. Before taking this decision, the researchers examined whether or not the coders' familiarity with the data was a factor, as one of the coders was not a member of the primary research team. From their account of the other study it is apparent that the researcher involved was also aware of her lack of knowledge of the context in which the data were collected and took steps to address this. In particular, she kept fieldnotes to enable her to reflect on the context of the secondary as well as the primary study. She also obtained further information on the context of the primary work by interviewing members of the primary research team about their experiences while working on the project, as well as reading their fieldnotes and listening to the tapes of their interviews (rather than relying on the transcripts alone).

Reflexivity was perhaps most strongly apparent in the reflections of Mauthner et al. (1998) on their respective efforts to re-use their own qualitative data. They experienced several problems in revisiting the data. As they explain:

> Most of our difficulties arose because we realized that our data were a product of the dynamic, dialectical and reflexive nature of a particular research encounter which both described but also delimited the meaning of the data. (Mauthner et al., 1998: 737)

In particular, they felt that with the passing of time they had become distanced from the material (although this was perceived to be an advantage by one of the authors). The personal and professional situation of the researchers had also changed over time, and was apparent in how their interest in aspects of the material had shifted. Finally, the variable involvement of the researchers in personally collecting the data also meant that memories of some interviews were stronger than others. For these reasons, the authors found the experience of revisiting their data to be problematic and to have far-reaching implications for the secondary analysis of archived data sets generated by other researchers, as well as the re-use of researchers' own data sets.

While I agree that reflexivity is a key issue for qualitative secondary analysis, I do not believe that it undermines the possibilities of re-using qualitative data. Indeed, it could be argued that Mauthner et al. (1998) place undue privilege on the author's initial analysis of the data while simultaneously playing down the context of this primary analysis. Reflexivity could therefore be used to highlight the respective contexts in which the data were originally collected and subject to secondary analysis in a different time and space by the same or by different researchers.

Conclusion

Clearly there is more to qualitative secondary analysis than analysis alone. The process of re-using qualitative data varies depending on whether data sets are obtained from personal, informal or formal sources but can involve: searching for relevant data sets; assessing the suitability of potential data sets; sorting the material; performing the secondary data analysis; assuring the quality of the work; carrying out related fieldwork (such as contacting the primary researcher(s), or collecting additional secondary or primary material to add to the data pot); and reporting the work. In addition, while the analysis of pre-existing data involves similar analytical approaches to those used in primary research, there was evidence to suggest that some approaches had been specially adapted for use in the context of secondary research (such as grounded theory) and that the analysis of secondary qualitative data can involve procedures which are not part of primary research (such as cleaning and re-coding raw data).

In general, however, the studies reviewed provided only limited information on how the above were accomplished and hence did not help to enhance practical knowledge of 'how to do' qualitative secondary analysis, nor facilitate appraisals of the quality of the work. I have therefore highlighted issues which researchers need to consider when assessing the re-usability of data sets and when reporting the results of their work. In addition, I have suggested that researchers could usefully develop more reflexive approaches to qualitative secondary analysis in order that the analysis can be situated in the various contexts to which it relates. Further research is required to document in more depth the methods and procedures for re-using qualitative data sets, including researchers' own personal data sets and data sets obtained through informal and formal data sharing.

Notes

1. Only a few authors briefly commented on the genesis of their studies (see Arksey and Sloper,1999; Bloor and McIntosh, 1990; Clarke-Steffen, 1998; Clayton et al., 1999; Robinson, 1993; Thorne, 1990a; Yamashita and Forsyth, 1998). For example, Yamashita and Forsyth's (1998) secondary study came about after the two investigators met while giving separate presentations at a conference and noted similarities in the themes arising from their respective primary work. Equally, researchers seldom discuss their decision to carry out research on a primary rather than a secondary basis. As a result, it is generally difficult to determine the exact reasons why particular studies were undertaken using secondary and/or primary methods.

2. Qualicat is available at http://www.qualidata.essex.ac.uk/search/qualicat.asp [accessed 28/3/2003].

3. Available at: http://lucy.ukc.ac.uk/Stirling/ [accessed 28/3/2003].

4. The DIPEx website is available at http://www.dipex.org/EXEC [accessed 28/3/2003].

5. Exceptions include Bloor (2000: 128) who examined the 'robustness of the indexing system' used to describe the data (which had been collected by other researchers). In their secondary study of Cohen and Taylor's (1972) research on long-term imprisonment, Fielding and Fielding (2000) also describe their assessment of the completeness of the data set.

6. See Corti and Ahmad (2000) for an account of the process of digitizing George Brown's team's research data, comprising twelve collections dating from 1969.

7. For example, in the United Kingdom, applicants to the Economic and Social Research Council can apply for funding to meet the costs of archiving the data set as part of the original research proposal.

8. For Thompson (2000a: para 41) the 'most valuable' data sets for secondary analysis are likely to have three qualities: a 'convincing' sample base; a relatively 'free-flowing' (open) interview format (exemplified by life stories); and where re-contact with the original informants is feasible, if required.

9. For overviews of these, see Bogdan and Biklen, 1992; Davies, 1999; Miles and Huberman, 1994; Miller and Dingwall, 1997; Murphy et al., 1998; Patton, 2002; Richardson, 1996; Silverman, 1993, 2000).

10. The latter technique of consulting with members who were not informants in the original research is an option for researchers who do not wish to make further demands on the original participants or who do not have access to them.

11. Audit trails are also potentially relevant to secondary researchers in two other ways: first, by enhancing the quality of documentation of primary research work and resulting meta-documentation of secondary data sets; secondly, by helping researchers to assess the re-usability of pre-existing data sets for use in proposed secondary research projects.

12. Similarly, electronic bibliographic databases do not yet have distinct keywords for classifying qualitative secondary studies (see Appendix A for information on the search strategies that were employed to identify these studies).

13. As an exception, Szabo and Strang (1999) provided a relatively full account of the conduct of the their secondary study, including information on: the respective funding of the primary and secondary work; the relationship of each of the authors to the data; how informed consent and ethical approval for the secondary analysis was obtained; the composition of the data set; the use of grounded theory in the analysis; how the secondary analysis was performed; how the data were managed; and how the rigor of the analysis was established.

14. For example, Thorne (1998) has suggested that, given their reliance on data from previous studies, secondary researchers need to account for any possible bias arising from the continued focus on the sample population concerned.

15. For an overview of available guidelines on reporting (and evaluating) qualitative research, see Reid and Gough (2000).

16. It should be noted that the British Sociological Association guidelines were not intended to cover all forms of qualitative research (Baxter, 2000).

17. Indeed, in some cases, it was difficult to distinguish whether a study was regarded as separate from, and secondary to, the primary work or, instead, one of multiple primary reports of different aspects of a single study.

7

The Future of Qualitative Secondary Analysis

The nature of qualitative secondary analysis	110
Secondary analysis and the 'landscape' of qualitative research	112
The secondary analyst as *bricoleur*	116
Resources for qualitative secondary analysis	119
Conclusion	124
Notes	124

In this concluding chapter I reflect on the main characteristics of qualitative secondary analysis and consider how it has been, and could be, developed as an approach in qualitative research. I begin by returning to the definition of secondary analysis postulated in Chapter 1 and considering, in the light of the review, whether it reflects the ways in which qualitative data have been re-used in practice. In the following section I examine the development of the methodology in the context of the historical 'landscape' of qualitative research outlined by Denzin and Lincoln (1994, 1998, 2000). I suggest that the debate surrounding the methodology exemplifies wider tensions in the relationship between quantitative and qualitative research, and between different traditions of qualitative research. I further suggest that the ways in which many exponents and proponents of qualitative secondary analysis have pioneered the methodology to date resembles the approach of the *'bricoleur'* in Denzin and Lincoln's account. Drawing on this metaphor, I explore the strengths and weaknesses of this style of qualitative research and consider its implications for the future development of qualitative secondary analysis. In conclusion, I highlight the resources that are needed to further develop and extend secondary analysis as a qualitative methodology.

The nature of qualitative secondary analysis

One of the objectives of this book was to shift from examining the potential of secondary analysis as a qualitative methodology to examining how it has actually been developed and used in practice. The following, provisional, definition of secondary analysis was used to inform a search for examples of qualitative secondary studies:

> Secondary analysis is a research strategy which makes use of pre-existing quantitative data or pre-existing qualitative research data for the purposes of investigating new questions or verifying previous studies.

The findings of the review of over 60 studies drawn mainly from the health and social care literature have shown that while the above definition conveys the possible range of functions of quantitative and qualitative secondary analysis, it does not capture the distinctive characteristics of qualitative secondary analysis. Four of the main distinguishing features of the methodology are summarized below.

First, as the above definition indicates, the concept of qualitative secondary analysis is more narrowly defined than that of quantitative secondary analysis. Whereas the latter refers to the re-use of statistical data derived from previous research *and* other contexts, the former is mainly concerned with the analysis of qualitative data from previous research studies. The focus on this type of 'non-naturalistic' data, such as interview data which have resulted from and been structured by the research process, distinguishes qualitative secondary analysis from the methodologies of documentary analysis and conversation analysis, which have been developed for the study of more 'naturalistic' types of qualitative data, such as autobiographies, photographs and speech. However, this distinction can be problematic in relation to some types of qualitative data, particularly life stories, which have been variously treated as both naturalistic and non-naturalistic data and hence may potentially be subject to documentary analysis or qualitative secondary analysis. Thus, in qualitative research, the distinction between research and non-research data, and therefore the scope of qualitative secondary analysis, is complicated by related definitions of what counts as non-naturalistic and naturalistic data.

Second, the review highlighted differences in the practice of qualitative and quantitative secondary analysis. For example, while qualitative and quantitative secondary analysis both involve the re-use of research data, these types of data tend to be drawn from different sources. Thus, while archived data sets are widely used in secondary studies published by quantitative researchers, most of the qualitative secondary studies that were reviewed were based on researchers' own previously collected data sets, or data which had been obtained through informal rather than formal data sharing. In addition, the review also revealed that while the above definition is correct in that the positivistic principle of verification is supported by some traditions of qualitative inquiry, such as conversation analysis, nonetheless in practice researchers have tended to re-use qualitative data for the purpose of doing new research and not in order to verify previous studies. Similarly, although alternative non-positivistic methods of externally and internally validating qualitative research have been developed for use in naturalistic inquiry, those which involve the retention and re-use of raw data, such as audit trails and referential adequacy, have not been fully or widely employed in practice.

Third, researchers were found to have developed particular approaches to re-using qualitative data in practice. Five types of secondary analysis were distinguished in Chapter 3, four of which were concerned with the investigation of new or additional questions. These varied in terms of the degree of divergence and convergence with the aims of the original study (supra and supplementary analysis), the use of multiple data sets (amplified analysis), the use of mixed primary and/or secondary sources of data (assorted analysis), and the use of data for the purposes of internal or external verification (re-analysis). Supplementary analysis was the most common type of study, used in 60 per cent of the studies reviewed. By contrast, only one example of re-analysis was found, and that involved the internal rather than external verification of the researcher's own previous analysis of a primary data set. Unlike the varieties of qualitative secondary analysis previously suggested by Thorne (1994, 1998), the types defined above were based on the review of actual rather than possible uses of the methodology.

Finally, it was found that the secondary analysis of qualitative data was generally perceived to be more problematic than that of quantitative data, and in comparison with the analysis of qualitative data performed in the context of primary research. This was particularly evident in the philosophical debates over the problems of data 'fit', not having 'been there', and verification (see Chapter 3), as well as the debates over the ethical and legal rights and responsibilities of research participants and researchers involved in primary and secondary qualitative research studies (see Chapter 5). While archivists and some researchers have suggested that there are methods for addressing these problems (see Corti and Thompson, 1998; Fink, 2000; Hinds et al., 1997; Szabo and Strang, 1997) and that these issues are not necessarily unique to qualitative secondary analysis (see Fielding and Fielding, 2000; Heaton, 1998), other researchers have queried whether the re-use of qualitative data is compatible with either the key tenets or ethical principles of qualitative inquiry (see Alderson, 1998; Mauthner et al., 1998; Vidich and Lyman, 1994). As I suggested in Chapter 4, the former perspective, held by various exponents and proponents of the methodology, exemplifies the 'technical' or 'methodological eclecticism' view of the relationship between qualitative and quantitative research, while the latter perspective, held by more sceptical researchers, is closer to the 'epistemological' or 'paradigm' view identified by Bryman (1988) and Hammersley (1997b) respectively.

In the next section I examine how the nature of qualitative secondary analysis, and the existence of these different views within qualitative research on the tenability of the methodology, relate to wider historical developments in qualitative research.

Secondary analysis and the 'landscape' of qualitative research

The emergence, establishment and decline of various traditions of qualitative inquiry in the twentieth century has been described in previous work (see Denzin, 1997; Denzin and Lincoln, 1994, 1995, 2000; Murphy et al., 1998). Denzin and Lincoln's (1994, 1995, 1998, 2000) depiction of the changing landscape of qualitative research is particularly relevant here because of its focus on North America where the potential of secondary analysis as a qualitative methodology was first recognized and where most of the studies reviewed were conducted. Although Denzin and Lincoln do not specifically discuss the example of qualitative secondary analysis, their work provides a useful context for analysing the makings of the methodology.

In the first edition of the *Handbook of Qualitative Research*, Denzin and Lincoln (1994) argue that qualitative research has passed through five phases or 'moments', and is about to enter a sixth (see also Denzin, 1997). They subsequently described this sixth moment, as well as the 'contours' of a seventh, in the second edition of the *Handbook* (Denzin and Lincoln, 2000). All seven moments are shown in Table 7.1.

The first 'traditional' moment spans the first half of the twentieth century and includes the Chicago school of sociology during its more 'positivistic' period. This was a time of empirical research documenting foreign cultures and marginal urban communities in the United States. In the second 'modernist' moment, which lasted from 1950 for two decades, qualitative methods such as grounded theory were developed and procedures for doing qualitative research were set out. This phase was succeeded by a period of 'blurred genres', when various theories and methods were in vogue and used in combination. Then, from the mid-1980s to 1990,

Table 7.1 The seven moments of qualitative research

1	1900–1950	Traditional
2	1950–1970	Modernist
3	1970–1986	Blurred genres
4	1986–1990	Crisis of representation
5	1990 1995	Post modern
6	1995–2000	Post-experimental inquiry
7	2000–	The future

Source: Based on Denzin and Lincoln (2000: 2–3; see also Denzin and Lincoln, 1994) with permission of Sage Publications.

there was a fourth moment characterized by a 'crisis of representation' and a related 'crisis of legitimation' when concerns were expressed about the ways in which researchers conventionally represented their inquiries and, in particular, their failure to reflect on their own part in the construction of this reality. In the fifth 'post-modern' moment of 1990–95, researchers turned from the grand theories and totalizing analyses of the past to more localized and provisional analyses of social life. The sixth moment of 'post-experimental inquiry' followed to 2000 and is characterized by political concerns. Finally, the seventh and 'future' moment is Denzin and Lincoln's initial outline and projection of the present, yet to be completed, phase of qualitative research.

This general historical shift from a secure, modernist conceptualization of qualitative research to a more vibrant and imaginative/creative post-modern view is, according to Denzin and Lincoln (1994, 1998, 2000), underpinned by two broad tensions which are a feature of the contemporary landscape of qualitative research. The first is the tension between past and present conceptualizations of qualitative inquiry which, for simplicity, I will broadly refer to as 'modern' and 'post-modern' forms of qualitative inquiry. These are divided by the crises of representation and legitimation when, as in Clifford and Marcus's (1986) *Writing Culture*, researchers began to question and problematize some of the hitherto taken for granted principles of qualitative research, particularly their own role in re-presenting the perspectives of others. The second and related set of tensions is between the various traditions of qualitative research which co-exist at present. They include a mix of modern and post-modern traditions drawing on different epistemologies (including positivism) and involving the use of a wide range of qualitative methods.

Denzin and Lincoln's analysis has been roundly criticised for focusing mainly on North America, creating arbitrary divisions, and for being progressive and skewed towards the present (Atkinson et al., 1999; Hammersley, 1999). However, it effectively plots the shifting and diverse nature of selected qualitative research traditions, the high points of these traditions, and the co-existence of particular traditions at different times. Like a canvas which has been over-painted from time to time, their landscape provides a basis for exploring how new methodologies, such as qualitative secondary analysis, emerge and develop over time, and how they relate to other methodologies of the same generation. Thus, one can trace the development of qualitative secondary analysis through this mutable landscape: from the 1960s, when its potential was first recognized by Glaser (1962, 1963), through the 1980s when a number of researchers began to carry out and publish qualitative secondary studies (particularly in North America), and the 1990s when researchers began to delineate the nature of the methodology and debate its potential, through to the beginning of the twenty-first century and the present focus on the use of the methodology in practice and how this might be developed in the future.

In this short history one can discern the ambivalence of researchers to the development of secondary analysis as a qualitative methodology. On the one hand, a growing number of researchers are re-using qualitative data in their work, some of whom have also promoted the potential of qualitative secondary analysis in related methodological work (Heaton, 1998; Hinds et al., 1997; Szabo and Strang, 1997; Sandelowski, 1997; Thorne, 1994, 1998), alongside various archivists and Qualidata (Corti, 1999, 2000; Corti et al., 1995; Corti and Thompson, 1998; Fink, 2000; James and Sørensen, 2000; Leh, 2000; Sheridan, 2000; Thompson, 2000a). Primary researchers are also increasingly donating their data sets to archives. On the other hand, as Qualidata's survey revealed, the research community as a whole includes researchers who support data archiving and data sharing in principle but not in practice, and researchers who have concerns about sharing and re-using qualitative data sets (Corti, 2000; Thompson, 2000a; see also Atkinson, 1998 and Jackson, 1990). Some of these researchers have published more sceptical or critical commentaries highlighting the difficulties and dangers of re-using qualitative data (Alderson, 1998; Hammersley, 1997a; Mauthner et al., 1998).

One can also observe the various co-existing traditions of qualitative inquiry that make up the contemporary landscape of qualitative research and examine their respective positions on the re-use of qualitative data. These include conversation analysis and naturalistic inquiry which, although they both promote the re-use of qualitative data, do so in ways which reflect their different epistemological positions. For example, as we saw in Chapter 4, these traditions of qualitative research have different views on the issue of verification: in conversation analysis, the publication of 'raw' data alongside analyses has been promoted as a means of enabling readers to independently scrutinize the findings whereas, in naturalistic inquiry, the concepts of the 'audit trail' and 'referential adequacy' have been recommended as ways of allowing the 'trustworthiness' of studies to be externally approved. The contemporary landscape of qualitative research also includes other traditions, such as observation-based ethnography, within which some researchers regard the re-use of archived qualitative data exemplified by the Human Relations Area Files (HRAF) as anathema (see Vidich and Lyman, 1994; Wax, 1997). Finally, as Denzin and Lincoln relate, this landscape also includes other post-modern approaches to qualitative research. These have not yet had an impact on qualitative secondary studies, although one secondary study has been published which illustrates the potential of more innovative and experimental work involving the re-use of qualitative data (Atkinson, 1992).

As noted above, exponents and proponents of qualitative secondary analysis tend to hold views which are consistent with the 'technical' or 'methodological eclecticism' view of qualitative research, while the views of more sceptical researchers are closer to those of the 'epistemological' or 'paradigm' position (Bryman, 1988; Hammersley, 1997b). Thus, qualitative secondary analysis has been most enthusiastically embraced by North American researchers using grounded theory and other techniques associated with the naturalistic inquiry tradition of qualitative research. They have developed various technical solutions to the problems of re-using qualitative data. For example, in response to the problem of data 'fit', researchers have adapted existing methods, such as theoretical sampling used in grounded theory, to facilitate the use of pre-existing data; they have also collected additional primary data to augment the secondary data. They have also minimized the problem of not having 'been there' by re-using their own data, or by involving the primary researchers who originally collected the data in the secondary analysis or at least consulting with them. Archivists have also contended that primary researchers can in any case minimize this difficulty by improving the meta-documentation of their data sets as a way of recovering as much of the context as possible. By contrast, the more cautious and sceptical researchers were from the United Kingdom and included self-styled 'interpretivists' who have expressed concern that the above technical fixes are inadequate and that pre-existing data could easily be mis-used by empiricist researchers who treat the data as 'given' and equally open to interpretation by primary and secondary researchers alike (Mauthner et al., 1998).

In the next section, I focus more closely on the exponents and proponents of qualitative secondary analysis and consider how their approach to re-using qualitative data compares with that of the qualitative researcher portrayed in Denzin and Lincoln's outline of the contemporary landscape of qualitative research.

The secondary analyst as *bricoleur*

Denzin and Lincoln claim that since the 'blurred genres' phase, qualitative researchers have become 'adept' at using various research strategies outlined in the *Handbook* (including case studies, ethnography, phenomenology and grounded theory) in an endeavour to interpret the social world (Denzin and Lincoln, 1995: 350; Denzin and Lincoln, 2000: xiii–xiv, 3). They compare this figure of the contemporary qualitative researcher to that of the *bricoleur* and the metaphor is expanded in their unfolding analysis of the landscape of qualitative research (Denzin and Lincoln, 1994, 1995, 1998, 2000; see also Denzin, 1994). In the first edition of the *Handbook*, they portray the *bricoleur* as someone who seeks solutions to problems using 'makeshift equipment, spare parts, and assemblage' (Denzin and Lincoln, 1994: 584), drawing on both the colloquial definition of the *bricoleur* as a 'Jack-of-all-trades' and Levi-Strauss's (1976) use of the term.[1] Elsewhere they defined the *bricoleur* as: 'One who uses any and all methods of inquiry to form better interpretations of ongoing social life' (Denzin and Lincoln, 1995: 349).

Later, in the second edition of the *Handbook* (Denzin and Lincoln, 2000: 6), they go on to specify five different types of *bricoleur*. These include: the 'interpretive *bricoleur* who 'understands that research is an interactive process shaped by his or her personal history, biography, gender, social class, race and ethnicity, and by those of the people in the setting'; the 'methodological *bricoleur*' who is able to perform various tasks; the 'theoretical *bricoleur*' who has a wide knowledge of a interpretive paradigms and perspectives and how these can be applied to particular problems; the 'narrative *bricoleur*' who is aware of the ways in which researchers from different paradigms construct their accounts of the social world; and the 'political *bricoleur*' who is aware of the power and impact of social research and that there is no 'value-free science'.

The relevance and value of this metaphor in qualitative research has been explored by Hammersley (1999) and McLeod (2000).[2] Hammersley (1999) criticizes Denzin and Lincoln for offering no alternatives to the metaphorical figure of the *bricoleur*, such as the engineer in Levi-Strauss' analysis of different modes of scientific thought, or the technician whom

Hammersley suggests as an example of a figure who follows a single method rigidly. However, Hammersley dismisses the *bricoleur* model altogether, preferring Neurath's metaphor of the scientist as a boat builder. In this model, ethnographers and qualitative researchers are compared to sailors who are forced to re-build their boat while still at sea, re-fashioning parts of the boat and at the same time maintaining sea worthiness. Thus, in contrast to the *bricoleur* model, where researchers' cobbling produces a *bricolage* which may be more or less cohesive, in the boat-building model it is vital that the product of researchers' labours fit together well – or else the boat will sink.

Whereas Hammersley dismisses the metaphor of the *bricoleur*, McLeod (2000) suggests that it is relevant to qualitative research in psychotherapy, where he claims researchers have acted as 'covert' *bricoleurs* because of pressures to conform to other models of qualitative research. In his view the metaphor is also positive because:

> creativity is one of the core characteristics of good qualitative research. It is very difficult to do *good* qualitative research, and those who do so achieve quality by drawing fully on their personal and cultural resources, rather than by following a set of laid-down procedures. The image and metaphor of the '*bricoleur*' captures well the resourcefulness of the qualitative researcher. (McLeod, 2000: conclusion. Original emphasis)

It is interesting in the light of this debate to explore to what extent the qualitative secondary analyst resembles the *bricoleur* as opposed to the alternative figures of the engineer, technician, or boat builder. Drawing on the findings of the review, the similarities and differences in the qualitative secondary analyst and the *bricoleur* are examined below.

The qualitative secondary analyst was found to resemble the *bricoleur* in three key respects. First, these researchers have made use of various existing resources in their work, including a wide range of data and, to a lesser extent, alternative theoretical perspectives and methods of analysis. The data that were re-used included: notes, transcripts and/or recordings of focus groups and interviews (including face to face interviews and telephone interviews); notes of participant observation; tapes of social interaction; vignettes; fieldnotes; letters; qualitative responses to open-ended questions in surveys; and unsolicited comments on questionnaires. This non-naturalistic secondary research data was analysed alongside other more naturalistic data, such as autobiographies, and other non-naturalistic primary qualitative research data in one type of secondary analysis (assorted analysis). Two or more data sets were drawn on in another type of secondary analysis (amplified analysis). The researchers also drew on different sources of pre-existing data – including their own personal data, and data sets obtained through informal data sharing and, to a lesser extent, from archives. In addition, alternative theoretical perspectives were used in some secondary studies (supra analysis) and alternative methods of analysis were used in one study in order to try and verify the results of the primary research (re-analysis).

Second, qualitative secondary analysts were found to be very adept at mining available resources in order to explore new or additional questions. In many cases, researchers used data from their personal primary research *oeuvre* for the purpose of investigating questions arising from their previous primary analysis of the same data set. They also adapted these resources where necessary, for example, sorting the data so that it fitted the purposes of the secondary analysis. Some researchers also drew on and revised methods originally developed for use in primary research, such as techniques used in grounded theory, to facilitate their secondary analysis. More generally, they have proposed, alongside archivists, various methodological innovations to facilitate qualitative secondary analysis, such as in the ethical protocols for obtaining informed consent and other permissions, in the procedures for documenting the process of primary qualitative research for archival purposes, and in the methods of preserving and sharing qualitative data sets.

Finally, the reports of qualitative secondary studies resembled a *bricolage* in the sense that they represented the culmination of a series of primary and secondary research activities by one or more researchers. However, as noted in Chapter 6, these reports generally focused on the process and findings of the secondary research, providing less information on the provenance of the data sets and the issues that were encountered in the course of re-using the material. In some cases, this resulted in a rather ambiguous *bricolage* where the boundaries between the primary and secondary research were blurred whereas, in other cases, the respective studies were set apart from each other.

While the metaphor generally befitted the exponents of qualitative secondary analysis, at the same time, it was found to apply to some researchers better than others. For example, qualitative secondary analysts were found to vary in their resourcefulness. While some carried out complex supra, amplified or assorted analyses using several data sets from different sources, the majority carried out relatively simple supplementary analyses of one of their own data sets. Similarly, only a few qualitative secondary analysts made use of archived data sets, or employed new theoretical perspectives, or alternative methods of analysis. In addition, some researchers who re-used qualitative data, particularly conversation analysts, did so in a relatively uncomplicated way. That is, they approached the use of pre-existing data in exactly the same ways as they used data from their primary research; indeed, they did not even distinguish between 'primary' and 'secondary' data or analysis. For these reasons, together with the fact that many qualitative researchers remain reluctant to re-use qualitative data, one may question whether, as Denzin and Lincoln imply, the metaphor applies to *all* qualitative researchers, 'There is no one way to do interpretive, qualitative inquiry. We are all interpretive *bricoleurs* stuck in the present working against the past as we move into the future' (Denzin and Lincoln, 2000: xv).

At the same time, the more innovative qualitative secondary analysts do strongly resemble the *bricoleur* in several respects. In their case, the metaphor is arguably more appropriate than, say, that of the engineer or the technician, because it captures the flexibility and eclecticism of their approach to qualitative research, while the other metaphors are more applicable to researchers who follow established protocols and rely on specialist tools and particular types of data in their work.

If one accepts that the metaphorical figures of the *bricoleur*, technician and other figures are a useful means of discriminating between different styles of qualitative research, each of which make a useful contribution to knowledge, this then shifts the emphasis from the question of whether it is wise for qualitative researchers in general to model themselves on one or other of these figures to the question of what do these different styles have to offer? It also raises related issues, such as: when is it (in)appropriate for qualitative researchers to act like *bricoleurs* and blur qualitative methods (Hammersley, 1999) and adopt 'creative' rather than 'conformist' approaches to research (McLeod, 2000)? What knowledge and expertise do researchers need to act as *bricoleurs* as opposed to technicians and so on?[3] As I suggest below, this depends not only on the judgement, experience and skills of individual analysts but also on the availability and quality of the resources that they have to work with.

Resources for qualitative secondary analysis

Throughout this book I have highlighted the various material, technical and intellectual resources which have already been developed and used by qualitative secondary analysts. I have also suggested areas where these resources could be made better use of, or are in need of further development. These are summarized in this final section in the hope that they might inform future research involving the collection, retention and/or re-use of qualitative research data for secondary purposes.

Material resources

Developments in data archiving and the promotion of data sharing have led to an increasing number of qualitative data sets being deposited and made available for re-use through data archives (see Chapter 2). The availability of archived qualitative data sets is, however, still limited compared to quantitative data sets; in particular, there is a relative lack of omnibus and longitudinal data sets that have been created to facilitate qualitative secondary studies. In addition, some of the qualitative data sets which have been archived to date, particularly the older data sets, were not created in the expectation that they would be archived and hence the quality of these data sets as secondary resources will depend on how well they have been retrospectively prepared as an archival resource. However, there are signs that growing awareness of the possibility of data archiving has begun to lead to changes in the ways in which primary studies are carried out, which should facilitate the archiving and re-use of qualitative research data in the future. These include changes in procedures for gaining informed consent, documenting the research, processing the data for deposit, and the increased involvement of archivists in the planning stages of primary research studies. Nevertheless, as Thompson (2000a) has pointed out, more qualitative studies could be created with a view to their utility as resources for secondary data analysis.

Qualitative data sets have also been more difficult to access and share than quantitative data sets because they have tended to be stored in paper form and on tape rather than electronically. Qualidata and some data archives have begun to explore the possibilities of digitizing data and sharing material on-line, which would potentially make access to these data a lot easier (Corti, 1999; Corti and Ahmad, 2000). In addition, in order to utilize archived data sets, researchers need to be able to identify, select and access relevant material. Again the activities of Qualidata and archives holding qualitative data sets have helped in this respect, particularly with the development of searchable on-line catalogues. However, these could be improved by, for example, providing more samples of the data linked to on-line catalogues and by asking data donors to highlight possible topics for future secondary investigation – or caveats against re-using the data in certain ways.

Although the lack of archived qualitative data sets and difficulties accessing the data *in situ* may have limited opportunities for using this source of qualitative data, secondary analysts have nevertheless made good use of other sources, including their own personal data sets and those obtained through informal data sharing with colleagues (who were usually involved in the secondary study as well). While this book has begun to explore the issues relating to the re-use of these different sources of data, there is a need for further research on the advantages and disadvantages of qualitative secondary analysis using auto-data and other research data obtained through informal and formal data sharing, as well as the actual procedures and techniques involved (see also 'Technical resources' below).

Technical resources

By 'technical' resources I mean the various tools that researchers can use, and practical knowledge that they can draw on, when collecting qualitative data and carrying out secondary studies. They include techniques and procedures developed or proposed by methodologists and archivists, as well as primary researchers and secondary qualitative researchers. As I have shown in Chapter 5, there are various useful guidelines for data donors on matters such as how to gain informed consent and request copyright transfers for qualitative secondary studies, and how to prepare qualitative data sets for deposit (Qualidata, undated a, b, c; Rolph, 2000; UK Data Archive, undated). However, there are relatively few guidelines for secondary data users on 'how to do' secondary analysis, with the exception of advice on how to assess the re-usability of data sets (see Hinds et al., 1997 and the revised criteria suggested in Chapter 6) and a few reflective accounts by exponents of the methodology which highlight some of the issues they encountered in the course of re-using qualitative data (Hinds et al., 1997; Mauthner et al., 1998; Szabo and Strang, 1997). Other, more general guidelines also exist and could be drawn on, such as the British Sociological Association Medical Sociology Group's (1996) proposed criteria for assessing the quality of qualitative research which, although they were designed for primary rather than secondary studies, are still broadly relevant and could be amended to include criteria that are more specific to the latter research (see Appendix B).

A good deal of the advice in the aforementioned guidelines has been reported and amplified in this book. However, there are a number of areas where knowledge is still lacking, particularly on 'how to do' qualitative secondary analysis. While this book has provided some insight into this process, it has been based on the limited information contained in published reports of qualitative secondary studies and a few commentaries on researchers' personal experiences of this process. Clearly there is a need to document in more detail what the process of doing qualitative secondary analysis involves in practice, including methods for re-using different types and sources of data for different purposes, and whether and how informed consent has been obtained for secondary studies. The development and implementation of methods for gaining informed consent would also be facilitated by research exploring research participants' views on the circumstances in which consent is and is not required, for what purposes, from whom, in what form, and at what point or points in the process of primary and secondary qualitative research it should be obtained.

I have also suggested that there may be a need to examine to what extent other techniques originally developed for use in the context of primary research are suitable for use in secondary research or can be adequately adapted for this purpose. This applies particularly to the techniques used in grounded theory and related procedures, such as the audit trail, which were originally developed for use in primary research. It also applies more generally to professional and ethical guidelines which vary in the extent to which they provide advice for researchers specifically on the principles and practices relating to qualitative secondary analysis. The review has also shown that some technical resources exist but have not been fully utilized in primary or secondary qualitative research. Further empirical research is required to examine whether techniques such as the audit trail and referential adequacy are being used in full but not reported in qualitative studies or, alternatively, whether qualitative researchers have improvised their use of these measures and, if so, why.[4]

Finally, under-reporting of the conduct of the secondary studies, including the provenance of the data sets, not only failed to inform researchers about what the process of doing qualitative secondary analysis involved in practice, but also limited the extent to which the quality of the secondary studies could be appraised. Ideally, standards of reporting qualitative secondary studies should be equivalent to those of primary qualitative research and should also reflect the unusual and intricate nature of these studies. This does not mean that reports have to be formulaic or duplicate information contained in primary research publications, but that key issues such as those set out in Table 6.2 in Chapter 6 are considered and reported where applicable.

Intellectual resources

The last set of intellectual resources encompasses knowledge of the nature of qualitative secondary analysis and the issues surrounding its use in principle and in practice, as well as awareness of the epistemological foundations of qualitative and quantitative methods in social research. It is hoped that this book has helped to clarify understanding of the concept of qualitative secondary analysis, showing how it differs from quantitative secondary analysis and is distinct from other qualitative and quantitative methodologies, such as documentary analysis, conversation analysis and meta-analysis. The typology outlined in Chapter 3 also serves to illustrate the various ways in which qualitative data have been re-used in practice in the area of health and social care research, just as Hyman's (1972) typology helped to discriminate between different emerging approaches to quantitative secondary analysis. An improved understanding of the complex and varied nature of the methodology should help qualitative researchers and bibliographers to better define and classify studies in ways which make it easier to identify and locate particular examples of the approach. The research for this book was made more difficult by the failure of authors and bibliographers to apply suitable key terms in electronic records of social research. Despite adopting additional strategies to locate relevant studies (see Appendix A) no doubt some examples were missed nonetheless.

More generally, it is hoped that the book has provided an insight into the epistemological tensions that exist within qualitative research, and between qualitative and quantitative research, and how these are manifest in the debates over the whys and wherefores of qualitative secondary analysis. Although exponents and proponents of the methodology have drawn on the intellectual resources of traditions such as naturalistic inquiry in their work, epistemological issues raised by the methodology were seldom discussed in the main reports of the studies. There was also a related lack of reflexivity in the studies, which is surprising given that the relationship between the secondary analyst and the data she or he uses is generally regarded as more problematic than in primary research. By ignoring the relationship between the researcher and the data, there is a danger that one of the key tenets of qualitative inquiry is broken and that the data are treated as 'given'. Thus, in general, qualitative secondary analysts could demonstrate more methodological and epistemological 'awareness' (Seale, 1999) in their work, and exercise more reflexivity in the main reports of their studies.

Finally, and related to the above, I have suggested that qualitative secondary analysts could perhaps begin to develop studies informed by newer, post-modern perspectives in qualitative research. For example, secondary analysis is potentially a useful approach for problematizing the ways in which qualitative data have been treated as 'given' and exploring the complex relationship between the analyst and the data – the *contexts of data interpretation* – as well as using the data to explore the perspectives of others – the *interpretation of the data in context*. This approach is to some extent compatible with forms of modernist qualitative research, such as subtle realism, which recognize the contingent nature of analyses. But it is perhaps more compatible with the turn to forms of post-modern qualitative research described by Denzin and Lincoln (1994, 1998, 2000), where the 'old' modernist concern with using particular scientific methods to access and interpret people's perspectives gives way to more eclectic, flexible and innovative styles of research. Thus, in post-modern qualitative research, primary researchers would collect and record stories from informants which they or archivists would then preserve for use in possible secondary studies. In using and re-using these data, primary and secondary researchers would work within, and move between, the various contexts in which the stories were told, recorded, rendered for analysis, interpreted and re-interpreted. A series of stories, each a *bricolage*, based on these primary and secondary data, would be produced and added to the resources available for qualitative researchers to work with.

Conclusion

In this final chapter and throughout the book I have suggested that secondary analysis does have the makings of a qualitative methodology, but that there are various epistemological, ethical and methodological issues arising from existing studies and commentaries which need to be considered and addressed in future studies and related theoretical and methodological work. I hope that this research has at least helped to identify the issues concerned and that it will inform future practice involving the collection, retention and re-use of qualitative research data. More generally, I hope that the book has conveyed something of the colourful and heterogeneous character of qualitative research and the myriad ways in which qualitative data has been, and could be, re-used within particular traditions of qualitative inquiry.

Notes

1. It should be noted that, as Hammersley (1999) points out, Denzin and Lincoln's use of the terms *bricoleur* and *bricolage* are equivocal and differ in important respects from that of Levi-Strauss (1976) and from colloquial usage. Thus, whereas Denzin and Lincoln regard the *bricoleur's* style as a model for social research, Levi-Strauss sees it as a mode of scientific reasoning to be studied by anthropologists. Denzin and Lincoln's (1994) suggestion that the *bricoleur* is an 'inventor' by virtue of having few tools or parts to work with also contrasts with Levi-Strauss's implied characterization of the *bricoleur* as someone who is more an improviser than an inventor, happily working within the limits of available resources. Finally, unlike colloquial uses, there is no suggestion of a lack of expertise in Denzin and Lincoln's account.
2. Both Hammersley's (1999) and McLeod's (2000) commentaries relate to Denzin and Lincoln's first use of the analogy in 1994.
3. Different views have been expressed on who should re-use qualitative data. On the one hand, the educational benefits of qualitative secondary analysis have been recognized (see Corti, 2000; Corti and Thompson, 1998; Szabo and Strang, 1997)

while, on the other hand, it has been suggested that it 'might be suited best to the skills of an experienced, rather than a novice, researcher' (Thorne, 1994: 272). In the review, secondary analysts varied in experience, ranging from doctoral students to established researchers.

4. For example, by keeping an audit trail but not actually involving an external auditor in the process.

Appendix A: The Literature Review

Background

The original literature review (Heaton, 2000) was conducted with funding from the Economic and Social Research Council (ESRC Award R000222918). It had three main aims:

- to describe the nature and use of secondary analysis of qualitative data sets in the international health and social care literature;
- to critically appraise these studies, in particular examining the epistemological, methodological and ethical issues arising from the use of secondary analysis in this area of research;
- to consider the overall quality and impact of the studies and the implications for future development of the methodology.

The review was conducted only a few years after Sally Thorne published the first paper on qualitative secondary analysis (Thorne, 1994) when it was not known to what extent the methodology had been used by qualitative researchers. One of the reasons for focusing on the international health and social care literature was that several of the methodological texts on the topic were written by researchers who worked in this area. Other reasons for focusing on this literature were as follows. First, studies in this area are relatively well covered by available electronic databases. Second, studies in this area employ a variety of qualitative methods, including interviews, focus groups and participant observation, providing a range of data for possible secondary research. Third, the varied nature of these studies was thought to enable the epistemological, ethical and methodological issues arising from the use of the methodology to be explored. Fourth, the policy and funding implications of making such data ready for secondary analysis and facilitating its re-use are significant given the amount of funding and number of studies conducted in this area. Finally, the review was limited to one major area of social research and not expanded to other areas because of the anticipated difficulties in

Table A1.1 Key terms used in the literature search

Qualitative	Revisit (and re-visit)
Qualitative method	Rework (and re-work)
Qualitative methodology/methodologies	Reexamine (and re-examine)
Qualitative data	Reexamination (and re-examination)
Qualitative reanalysis (and re-analysis)	Reanalyze (and re-analyze, reanalyse, re analyse)
Secondary	
Secondary analysis	Restudy (and re-study)
Secondary interview data	Restudies (and re-studies)
Secondary qualitative data	Repeat study/studies
Post hoc	Replication
Retrospective	Verification
Data sharing	Corroborate/corroboration
Archived data	Documentary analysis
Data archives	Life history
Data repositories	Oral history
Reuse (and re-use)	Autobiography

identifying examples of secondary analysis, which are not always clearly defined as such (Heaton, 2000). Thus, electronic searches had to be supplemented by other more time-consuming manual strategies which would have greatly increased the duration of the review if it had been extended to other areas. Where appropriate, however, the wider social research literature was searched for examples of secondary studies with specific characteristics. For example, an effort was made to identify re-analyses and re-studies in other areas because of the apparent lack of such studies in the health and social care literature. In addition, any examples of secondary studies identified from outside the health and social care literature were appraised to see if they shared the characteristics of the studies included in the main review.

Conduct of the review

A number of resources were searched (electronically and by hand) to identify examples of secondary studies of qualitative data. These included searches of studies recorded on electronic databases from 1980 onwards (including BIDS Social Science Citation Index, CINAHL, EMBASE, Medline and Sociological Abstracts); search engines (Google, AltaVista); and publishers' electronic bibliographies (Sage, Cambridge University Press). Hand searches of key journals (such as *Qualitative Health Research*), study references and key author citations were also carried out. In addition, appeals for information were issued via the British Sociological

Table A1.2 Aspects of studies appraised in the review

Author(s) of the secondary study.
Study reference number (for the review).
Year of related primary research publication.
Year of secondary study publication.
Place of publication (e.g. journal, chapter details).
Country of origin of the secondary study (i.e. research setting or first author's base).
Context of primary study (e.g. doctoral research, commissioned project).
Primary study funder/award reference number (if stated in acknowledgements).
Context of secondary research (e.g. doctoral research, commissioned project).
Secondary study funder/award reference number (if stated in acknowledgements).
Relationship of the secondary researcher(s) to the primary researcher(s) (i.e. independent, semi-independent or non-independent, depending on whose data set(s) was re-used).
Whether or not use of 'secondary analysis' was defined as such? (and, if not, how the study methodology was alternatively defined).
Purpose of the primary study.
Purpose of the secondary study.
Number of data sets used in the secondary study.
Type(s) of qualitative data used in the secondary study (e.g. from previous research or from naturalistic administrative records).
Type of research data used in the secondary study (e.g. interview tapes and transcripts, focus group tapes and transcripts).
Size of the sample in the primary study from which the data were derived.
Size of the sample in the secondary study.
Methods and quality assurance strategies used in the secondary study.
Whether or not the secondary analysis was computer-assisted.
Whether or not informed consent for the secondary analysis was obtained. (And, if so, how).
Whether any additional (primary or secondary naturalistic) data was collected for and used in the secondary study.
Links between secondary studies (e.g. multiple secondary studies using the same data set(s)).
Comments (e.g. special features of the study).

Association's *Network* magazine, conference flyers, electronic discussion lists (QUALS-R and ARCHIVE-QUALITATIVE-DATA), and personal contacts. As the electronic databases did not systematically key-word the use of the methodology, a number of key terms were derived from the studies initially identified (see Table A1.1). These were used to systematically search study titles and abstracts for examples of relevant studies.

Abstracts of studies identified by these means were read and the full papers were obtained if they met the inclusion criteria – that the study involved the re-use of qualitative data in the area of health and social care, and was published in English. Studies which were clearly secondary analyses of quantitative data or meta-analyses were excluded; studies from other areas were also excluded from the original review (after being

Table A1.3 Principal characteristics of types of secondary analysis

Type	Function	Focus	Type of data	Source(s) of Secondary data
Supra analysis	To investigate a new research question or issue	Transcends primary study	Secondary data set(s) *or* mixed secondary and primary data	Auto-data *or* independent data
Supplementary analysis	To investigate a new research question	In-depth analysis of an aspect of the primary study	Secondary data set(s) *or* mixed secondary and primary data	Auto-data *or* independent data
Re-analysis	To re-investigate primary study question	Same focus as primary study	Secondary data set	Auto-data *or* independent data
Amplified analysis	To investigate a new research question	Transcends primary study *or* in-depth analysis of an aspect of the primary studies	Multiple secondary data sets	Auto-data *and/or* independent data
Assorted analysis	To investigate a new research question	Transcends primary study *or* in-depth analysis of an aspect of the primary study	Mixed secondary and primary data sets	Auto-data *and/or* independent data

briefly appraised for comparative purposes). It was not always possible from reading the abstract to determine whether the secondary analysis was of quantitative or qualitative data, nor whether the data involved were collected as part of a previous research study or for the study at hand. This was particularly the case for studies which utilized life stories.

The resulting studies were read and information extracted and entered on to a spreadsheet for reference (see Table A1.2). Bibliographic details and a short outline of each study were also entered onto a computer using EndNote. This was used to produce an annotated bibliography of the earlier findings (Heaton, 2000). In the detailed appraisal of the studies, five key aspects of the studies were examined:

- *function* – whether or not the studies were designed to investigate new research issues or to verify previous analyses;
- *focus* – how the question(s) addressed by those secondary studies investigating new research questions differed from the aims of the primary research;

- *methods/perspectives* – whether or not these were the same as those employed in the primary research;
- *range of data set(s)* – whether one or more secondary data sets were used and whether any primary research was also conducted alongside the secondary analysis;
- *origins of data set(s)* – the extent to which studies relied on auto-data and/or data from other sources.

These dimensions were initially defined *a priori*, on the basis of previous conceptualizations of the methodology, and then refined as the review progressed to ensure that they reflected the main variations in the literature. Based on this information, six varieties of secondary analysis were initially discerned: supra analysis, amplified analysis, supplementary analysis, complementary analysis (which I have referred to as assorted analysis in this text), alternative analysis and repeat analysis. Given the small number of alternative and repeat analyses these have been collapsed into one category – re-analysis (see Table A1.3). Further research is required to establish the extent to which qualitative studies have been verified in social research and the ways in which this has been approached.

Appendix B: Criteria for the Evaluation of Qualitative Research Papers

British Sociological Association Medical Sociology Group guidelines

1. Are the methods of the research appropriate to the nature of the question being asked? i.e.
   - Does the research seek to understand the processes or structures, or illuminate subjective experiences or meanings?
   - Are the categories or groups being examined of a type which cannot be preselected, or the possible outcomes cannot be specified in advance?
   - Could a quantitative approach have addressed the issue better?
2. Is the connection to an existing body of knowledge or theory clear? i.e.
   - Is there adequate reference to the literature?
   - Does the work cohere with, or critically address, existing theory?

Methods

3. Are there clear accounts of the criteria used for the selection of subjects for study, and of the data collection and analysis?

4. Is the selection of cases or participants theoretically justified?
   - The unit of research may be people, or events, institutions, samples of natural behaviour, conversations, written material etc. In any case, while random sampling may not be appropriate, is it nevertheless clear what population the sample refers to?
   - Is consideration given to whether the units chosen were unusual in some important way?

5. Does the sensitivity of the methods match the needs of the research questions?
   - Does the method accept the implications of an approach which respects the perceptions of those being studied?

- To what extent are any definitions or agendas taken for granted, rather than being critically examined or left open?
- Are the limitations of any structured interview method considered?

6. Has the relationship between fieldworkers and subjects been considered, and is there evidence that the research was presented and explained to its subjects?
    - If more than one worker was involved, has comparability been considered?
    - Is there evidence about how the subjects perceived the research?
    - Is there evidence about how any group processes were conducted?
7. Was the data collection and record keeping systematic? e.g.
    - Were careful records kept?
    - Is there evidence available for independent examination?
    - Were full records or transcripts of conversations used if appropriate?

Analysis

8. Is reference made to accepted procedures for analysis?
    - Is it clear how the analysis is done? (Detailed repetition of how to perform standard procedures ought not to be expected).
    - Has its reliability been considered, ideally by independent repetition?
9. How systematic is the analysis?
    - What steps were taken to guard against selectivity in the use of data?
    - In research with individuals, is it clear that there has not been selection in some cases and ignoring of less interesting ones? In group research, are all categories of opinion taken into account?
10. Is there adequate discussion of how themes, concepts and categories were derived from the data?
    - It is sometimes inevitable that externally given or predetermined descriptive categories are used, but have they been examined for their real meaning or any possible ambiguities?
11. Is there adequate discussion of the evidence both for and against the researcher's arguments?
    - Is negative data given? Has there been any search for cases which might refute the conclusions?
12. Have measures been taken to test the validity of the findings?
    - For instance, have methods such as feeding them back to the respondents, triangulation, or procedures such as grounded theory been used?

13. Have any steps been taken to see whether the analysis would be comprehensible to the participants, if this is possible and relevant?
    - Has the meaning of their accounts been explored with respondents?
    - Have apparent anomalies and contradictions been discussed with them, rather than assumptions being made?

Presentation

14. Is the research clearly contextualized?
    - Is all the relevant information about the setting and the subjects supplied?
    - Are the cases or variables which are being studied integrated in their social context, rather than being abstracted and decontextualized?
15. Are the data presented systematically?
    - Are quotations, fieldnotes, etc. identified in a way which enables the reader to judge the range of evidence being used?
16. Is a clear distinction made between the data and its interpretation?
    - Do the conclusions follow from the data? (It should be noted that the phases of research – data collection, analysis, discussion – are not usually separate and papers do not necessarily follow the quantitative pattern of methods, results, discussion).
17. Is sufficient of the original evidence presented to satisfy the reader of the relationship between the evidence and the conclusions?
    - Though the presentation of discursive data is always going to require more space than numerical data, is the paper as concise as possible?
18. Is the author's own position clearly stated?
    - Is the researcher's perspective described?
    - Has the researcher examined their own role, possible bias, and influence on the research?
19. Are the results credible and appropriate?
    - Do they address the research question(s)?
    - Are they plausible and coherent?
    - Are they important, either theoretically or practically, or trivial?

Ethics

20. Have ethical issues been adequately considered?
    - Is the issue of confidentiality (often particularly difficult in qualitative work) been adequately dealt with?

- Have the consequences of the research – including establishing relationships with the subjects, raising expectations, changing behaviour, etc. – been considered?

Source: Reprinted from British Sociological Association Medical Sociology Group (1996: 35–7) with permission of BSA Publications Limited, a subsidiary of the British Sociological Association.

Elliot, Fischer and Rennie (1994) guidelines for assessing the quality of qualitative research

[(1) Manuscripts be] of archival significance. That is ... [they] contribute to the building of the discipline's body of knowledge and understanding. (2) The manuscript specifies where the study fits within relevant literature and indicates the intended contributions (purposes or questions) of the study. (3) The procedures used are appropriate or responsive to the intended contributions (purposes of questions posed for the study). (4) Procedures are specified clearly so that readers may see how to conduct a similar study themselves and may judge for themselves how well the study followed its stated procedures. (5) The research results are discussed in terms of their contribution to theory, content, method, and/or practical domains. (6) Limitations of the study are discussed. (7) The manuscript is written clearly, and any necessary technical terms are defined. (8) Any speculation is clearly identified as speculation. (9) The manuscript is acceptable to reviewers familiar with its content area and with its method(s).

Source: Elliot et al. (1994). Quoted in Lincoln, 1995: 279.

Bibliography

Abel, E. and Sherman, J.J. (1991) 'Use of national data sets to teach graduate students research skills', *Western Journal of Nursing Research*, 13 (6): 794–97.

Adams, A., Hardey, M. and Mulhall, A. (1994) 'Secondary analysis in nursing research', in Hardey, M. and Mulhall, A. (eds), *Nursing Research: Theory and Practice*. London: Chapman & Hall.

Aita, V.A. and Crabtree, B. (2000) 'Historical reflections on current preventive practice', *Preventive Medicine*, 30: 5–16.

Alderson, P. (1998) 'Confidentiality and consent in qualitative research', *Network – Newsletter of the British Sociological Association*, (69): 6–7.

Alderson, P. (2001) 'On doing qualitative research linked to ethical healthcare. Volume 1'. London: The Wellcome Trust.

Allen and Overy (1998) 'Copyright/confidentiality: final report to the Economic and Social Research Council.' Retrieved from: ftp://ftp.esrc.ac.uk/pub/guide.doc [accessed 14/9/1998].

American Sociological Association (1997) 'American Sociological Association Code of Ethics.' Available at http://www.asanet.org/members/ecoderev.html [accessed 10/2/2003].

Angell, R. (1945) 'A critical review of the development of the personal document method in Sociology 1920–1940', in Gottschalk, L., Kluckholn, C. and Angell, R., *The Use of Personal Documents in History, Anthropology, and Sociology*. New York: Social Science Research Council. pp. 175–232.

Angst, D.B. and Deatrick, J.A. (1996) 'Involvement in health care decisions: parents and children with chronic illness', *Journal of Family Nursing*, 2 (2): 174–94.

Arksey, H. and Sloper, P. (1999) 'Disputed diagnoses: the cases of RSI and childhood cancer', *Social Science & Medicine*, 49: 483–97.

Ashmore, M. and Reed, D. (2000) 'Innocence and nostalgia in conversation analysis: the dynamic relations of tape and transcript', *Forum: Qualitative Social Research* [Online Journal], 1 (3): 45 paragraphs. Available at http://qualitative-research.net/fqs/fqs-eng.htm [accessed 28/8/2003].

Association of Social Anthropologists of the UK and the Commonwealth (1999) 'Ethical guidelines for good research practice.' Available at http://les1.man.ac.uk/asa/Ethics/ethics.htm [accessed 10/2/2003].

Atkinson, J.M. and Heritage, J. (eds) (1984) *Structures of Social Action: Studies in Conversation Analysis*. Cambridge: Cambridge University Press.

Atkinson, P. (1992) 'The ethnography of a medical setting: reading, writing and rhetoric', *Qualitative Health Research*, 2 (4): 451–74.

Atkinson, P., Coffey, A. and Delamont, S. (1999) 'Ethnography: post, past, and present', *Journal of Contemporary Ethnography*, 28 (5): 460–71.

Atkinson, R. (1998) *The Life Story Interview*. London: Sage.

Australian National Health and Medical Research Council (1999) 'National statement on ethical conduct in research involving humans.' Available at http://www.health.gov.au/nhmrc/publications/pdf/e35.pdf [accessed 29/04/03].

Backett, K.C. and Davison, C. (1995) 'Lifecourse and lifestyle: the social and cultural location of health behaviours', *Social Science and Medicine*, 40 (5): 629–38.

Backhouse, G. (2002) 'How preserving confidentiality in qualitative health research can be compatible with preserving data for future use', *Medical Sociology News*, 28 (3): 32–35.

Barnes, J. (1979) *Who Should Know What? Social Science, Privacy and Ethics*. Cambridge: Cambridge University Press.

Baron, J.N. (1988) 'Data sharing as a public good', *American Sociological Review*, 53 (1): vi–viii.

Baxter, M. (2000) 'Criteria for qualitative research', *Medical Sociology News*, 26 (2): 34–37.

Beck, C. (1993) 'Qualitative research: the evaluation of its credibility, fittingness, and auditability', *Western Journal of Nursing Research*, 15 (2): 263–66.

Bevan, M. (2000) 'Family and vocation: career choice and the life histories of general practitioners', in Bornat, J., Perks, R., Thompson, P. and Walmsley, J. (eds), *Oral History, Health and Welfare*. London: Routledge. pp. 21–47.

Bloor, M. (1997) 'Techniques of validation in qualitative research: a critical commentary', in Miller, G. and Dingwall, R. (eds), *Context and Method in Qualitative Research*. London: Sage. pp. 37–50.

Bloor, M. (2000) 'The South Wales Miners Federation, Miners' Lung and the instrumental use of expertise, 1900–1950', *Social Studies in Science*, 30 (1): 125–40.

Bloor, M. and McIntosh, J. (1990) 'Surveillance and concealment: a comparison of techniques of client resistance in therapeutic communities and health visiting', in Cunningham-Burley, S. and McKeganey, N.P. (eds), *Readings in Medical Sociology*. London: Tavistock/Routledge. pp. 159–81.

Bloor, M., McKeganey, N. and Fonkert, D. (1988) *One Foot in Eden: A Sociological Study of the Range of Therapeutic Community Practice*. London: Routledge.

Boddy, M. (2001) 'Data policy and data archiving: report on consultation for the ESRC Research Resources Board'. Bristol: University of Bristol.

Boelen, W.A.M. (1992) 'Street Corner Society: Cornerville revisited', *Journal of Contemporary Ethnography*, 21 (1): 11–51.

Bogdan, R.C. and Biklen, S.K. (1992) *Qualitative Research for Education: An Introduction to Theory and Methods*. Boston: Allyn and Bacon.

Bond, G.C. (1990) 'Fieldnotes: research in past occurrences', in Sanjek, R. (ed.) *Fieldnotes: The Makings of Anthropology*. Ithaca: Cornell University Press. pp. 273–89.

Boruch, R.F. and Cordray, D.S. (1985) 'Professional codes and guidelines in data sharing', in Fienberg, S.E., Martin, M.E. and Straf, M.L. (eds), *Sharing Research Data*. Washington DC: National Academy Press. pp. 199–223.

Breckenridge, D.M. (1997) 'Decisions regarding dialysis treatment modality: a holistic perspective', *Holistic Nursing Practice*, 12 (1): 54–61.

Bristol Royal Infirmary Report (2001) *Learning from Bristol: the report of the public inquiry into children's heart surgery at the Bristol Royal Infirmary 1984–1995*. Command Paper: CM 5207. Available at: http://www.bristol-inquiry.org.uk/index.htm [accessed 11/8/2003].

British Sociological Association (undated, c. 2002) 'Statement of ethical practice'. Available at http://www.britsoc.co.uk/bsaweb.php?link_id=14&area=item1 [accessed 10/2/2003].

British Sociological Association Medical Sociology Group (1996) 'Criteria for the Evaluation of Qualitative Research Papers', *Medical Sociology News*, 22 (1): 69–71.
Brown, R.B. and Brooks, I. (2002) 'The temporal landscape of night nursing', *Journal of Advanced Nursing*, 39 (4): 384–90.
Bryman, A. (1988) *Quantity and Quality in Social Research*. London: Unwin Hyman.
Bull, M.J. and Kane, R.L. (1996) 'Gaps in discharge planning', *The Journal of Applied Gerontology*, 15 (4): 486–500.
Bulmer, M. (ed.) (1979) *Censuses, Surveys and Privacy*. London: Macmillan.
Burgess, R.G. (1981) 'An Ethnographic Study of a Comprehensive School'. Unpublished thesis: University of Warwick.
Burgess, R.G. (1983) *Experiencing Comprehensive Education: A Study of Bishop McGregor School*. London: Methuen.
Burgess, R.G. (1987) 'Studying and restudying Bishop McGregor School', in Walford, G. (ed.) *Doing Sociology of Education*. London: The Falmer Press. pp. 67–94.
Burns, N. and Grove, S. (1993) *The Practice of Nursing Research: Conduct, Critique and Utilization*. Philadelphia: WB Saunders.
Burstein, L. (1978) 'Secondary analysis: an important resource for educational research and evaluation', *Educational Researcher*, 7 (5): 9–12.
Calder, A. and Sheridan, D. (eds) (1984) *Speak For Yourself. A Mass Observation Anthology, 1937–49*. London: Jonathon Cape.
Campbell, D. and Fiske, D. (1959) 'Convergent and discriminant validation by multitrait multi-dimensional matrix', *Psychological Bulletin*, 56: 81–105.
Canadian Sociology and Anthropology Association (1994) 'Statement of Professional Ethics.' Available at http://alcor.concordia.ca/~csaa1/index.html [accessed 10/2/2003].
Caplow, T., Bahr, H., Chadwick, B., Hill, R. and Williamson, M. (1982) *Middletown Families: Fifty Years of Change and Continuity*. Minneapolis: University of Minnesota Press.
Caplow, T., Bahr, H., Chadwick, B., Hill, R. and Williamson, M. (1983) *All Faithful People: Change and Continuity in Middletown's Religion*. Minneapolis: University of Minnesota Press.
Ceci, S.J. and Walker, E. (1983) 'Private archives and public needs', *American Psychologist*, 38: 414–23.
Cecil, J.S. and Boruch, R. (1988) 'Compelled disclosure of research data: an early warning and suggestions for psychologists', *Law and Human Behaviour*, 12 (2): 181–89.
Cherlin, A. (1991) 'On analysing other people's data', *Developmental Psychology*, 27 (6): 946–48.
Clarke-Steffen, L. (1998) 'The meaning of peak and nadir experiences of pediatric oncology nurses: secondary analysis', *Journal of Pediatric Oncology Nursing*, 15 (1): 25–33.
Clayton, D.L.K., Rogers, S. and Stuifbergen, A. (1999) 'Answers to unasked questions: writing in the margins', *Research in Nursing & Health*, 22: 512–22.
Clifford, J. and Marcus, G. (eds) (1986) *Writing Culture*. Berkeley: University of California Press.
Clubb, J.M., Austin, E.W., Geda, C.L. and Traugott, M.W. (1985) 'Sharing research data in the Social Sciences', in Fienberg, S.E., Martin, M.E. and Staff, M.L. (eds), *Sharing Research Data*. Washington DC: National Academy Press. pp. 39–88.
Cohen, M.H. (1995) 'The triggers of heightened parental uncertainty in chronic, life-threatening childhood illness', *Qualitative Health Research*, 5 (1): 63–77.

Cohen, S. and Taylor, L. (1972) *Psychological Survival: The Effects of Long-Term Imprisonment*. London: Allen Lane.

Colby, A. (1982) 'The use of secondary analysis in the study of women and social change', *Journal of Social Issues*, 38 (1): 119–23.

Colby, A. and Phelps, E. (1990) 'Archiving longitudinal data', in Magnusson, D. and Bergman, L.R. (eds), *Data Quality In Longitudinal Research*. Cambridge: Cambridge University Press. pp. 249–62.

Connor, S.S. and Fuenzalida-Puelma, H.L. (eds) (1990) *Bioethics: Issues and Perspectives*. Washington DC: Pan American Health Organization.

Corti, L. (1999) 'Text, sound and videotape: the future of qualitative data in the global network', *Newsletter – International Association for Social Science Information Service and Technology* (IASSIST), 23 (2): 18–25.

Corti, L. (2000) 'Progress and problems of preserving and providing access to qualitative data for social research – the international picture of an emerging culture', *Forum: Qualitative Social Research* [Online Journal], 1 (3): 58 paragraphs. Available at http://qualitative-research.net/fqs/fqs-eng.htm [accessed 28/8/2003].

Corti, L. and Ahmad, N. (2000) 'Digitising and providing access to socio-medical case records: the case of George Brown's works', *Forum: Qualitative Social Research* [Online Journal], 1 (3): 19 paragraphs. Available at http://qualitative-research.net/fqs/fqs-eng.htm [accessed 28/8/2003].

Corti, L., Day, A. and Backhouse, G. (2000) 'Confidentiality and informed consent: issues for consideration in the preservation of and provision of access to qualitative data archives', *Forum: Qualitative Social Research* [Online Journal], 1 (3): 46 paragraphs. Available at http://qualitative-research.net/fqs/fqs-eng.htm [accessed 28/8/2003].

Corti, L., Foster, J. and Thompson, P. (1995) 'Archiving qualitative research data', *Social Research Update*, (10).

Corti, L. and Thompson, P. (1998) 'Are you sitting on your qualitative data? Qualidata's mission', *International Journal of Social Research Methodology*, 1 (1): 85–89.

Creswell, J.W. (1998) *Qualitative Inquiry and Research Design: Choosing Among Five Traditions*. Thousand Oaks: Sage.

Dale, A., Arber, S. and Procter, M. (1988) *Doing Secondary Analysis*. London: Unwin Hyman.

Davies, C.A. (1999) *Reflexive Ethnography: A Guide to Researching Selves and Others*. London: Routledge.

Deatrick, J.A., Knafl, K.A. and Guyer, K. (1992) 'The meaning of caregiving behaviours: inductive approaches to family theory development', in Feetham, S.L., Meister, S.B., Bell, J.M. and Gilliss, C.L. (eds), *The Nursing of Families: Theory/Research/Education/Practice*. Newbury Park: Sage. pp. 38–45.

Denzin, N.K. (1970) *The Research Act in Sociology: A Theoretical Introduction to Sociological Methods*. London: Butterworths.

Denzin, N.K. (1992) 'Whose Cornerville is it, anyway?', *Journal of Contemporary Ethnography*, 21 (2): 120–32.

Denzin, N.K. (1994) 'Romancing the text: the qualitative researcher-writer-as-bricoleur', *Bulletin of the Council for Research in Music Education*, 22 (Fall): 15–30.

Denzin, N.K. (1997) *Interpretive Ethnography: Ethnographic Practices for the 21st Century*. Thousand Oaks: Sage.

Denzin, N.K. and Lincoln, Y.S. (eds) (1994) *Handbook of Qualitative Research*. London: Sage. (First edition).

Denzin, N.K. and Lincoln, Y.S. (1995) 'Transforming qualitative research methods: is it a revolution?', *Journal of Contemporary Ethnography*, 24 (3): 349–58.

Denzin, N.K. and Lincoln, Y.S. (eds) (1998) *The Landscape of Qualitative Research: Theories and Issues*. Thousand Oaks: Sage.

Denzin, N.K. and Lincoln, Y.S. (eds) (2000) *Handbook of Qualitative Research*. London: Sage. (Second edition).

Devine, F. (1990) 'Privatism and the working class: affluent workers in the 1980s?'. Colchester: University of Essex.

Devine, F. (1992) 'Social identities, class identity and political perspectives', *The Sociological Review*: 229–52.

Dixon-Woods, M., Fitzpatrick, R. and Roberts, K. (2001) 'Including qualitative research in systematic reviews: problems and opportunities', *Journal of Evaluation in Clinical Practice*, 7: 125–33.

Doherty, W.J., Lester, M.E. and Leigh, G. (1986) 'Marriage encounter weekends: couples who win and couples who lose', *Journal of Marital and Family Therapy*, 12 (1): 49–61.

Dunn, C.S. and Austin, E.W. (1998) 'Protecting confidentiality in archival data resources', *Newsletter – International Association for Social Science Information Service and Technology* (IASSIST), 22 (2): 16–24.

Durkheim, E. (1952) *Suicide: A Study in Sociology*. London: Routledge and Kegan Paul. (First published 1897).

Earle, J.R., Knudsen, D.D. and Shriver, D.W. (1976) *Spindles and Spires: A Re-Study of Religion and Social Change in Gastonia*. Atlanta: John Knox Press.

Economic and Social Research Council (undated, c. 1998) 'Guidelines on copyright and confidentiality: legal issues for social science researchers'. Available at: http://www.esrc.ac.uk/esrccontent/DownloadDocs/wwwcopyrightandconfidentiality.doc [accessed 28/8/2003].

Eisner, E.W. (1975) 'The perceptive eye: toward the reformulation of educational evaluation', *Occasional Papers of the Stanford Evaluation Consortium*. Stanford, CA: Stanford University.

Elder, G.H., Pavalko, E.K. and Clipp, E.C. (1993) *Working With Archival Data: Studying Lives*. London: Sage.

Elliot, R., Fischer, C. and Rennie, D. (1994) 'Evolving guidelines for publication of qualitative research studies'. Unpublished manuscript. (Discussed by Lincoln, 1995).

Erickson, K. and Stull, D. (1998) *Doing Team Ethnography: Warnings and Advice*. Thousand Oaks: Sage.

Estabrooks, C.A., Field, P.A. and Morse, J.M. (1994) 'Aggregating qualitative findings: an approach to theory development', *Qualitative Health Research*, 4 (4): 503–11.

Estabrooks, C.A. and Romyn, D.M. (1995) 'Data sharing in nursing research: advantages and challenges', *Canadian Journal of Nursing Research*, 27 (1): 77–88.

Fendrich, M., Wislar, J.S., Mackesy-Amiti, M.E. and Goldstein, P. (1996) 'Mechanisms of noncompletion in ethnographic research on drugs: results from a secondary analysis', *Journal of Drug Issues*, 26 (1): 23–44.

Ferrell, B.R., Grant, M., Dean, G.E., Funk, B. and Ly, J. (1996) '"Bone tired": the experience of fatigue and its impact on quality of life', *Oncology Nursing Forum*, 23 (10): 1539–47.

Fielding, N. (2000) 'The shared fate of two innovations in qualitative methodology: the relationship of qualitative software and secondary analysis of archived qualitative data', *Forum: Qualitative Social Research* [Online Journal], 1 (3): 43 paragraphs. Available at http://qualitative-research.net/fqs/fqs-eng.htm [accessed 28/8/2003].

Fielding, N.G. and Fielding, J.L. (2000) 'Resistance and adaptation to criminal identity: using secondary analysis to evaluate classic studies of crime and deviance', *Sociology*, 34 (4): 671–89.

Fienberg, S.E., Martin, M.E. and Staff, M.L. (1985) *Sharing Research Data*. Washington DC: National Academy Press.

Fink, A.S. (2000) 'The role of the researcher in the qualitative research process. A potential barrier to archiving qualitative data', *Forum: Qualitative Social Research* [Online Journal], 1 (3): 69 paragraphs. Available at http://qualitative-research.net/fqs/fqs-eng.htm [accessed 28/8/2003].

Freeman, D. (1983) *Margaret Mead and Samoa: The Making and Unmaking of an Anthropological Myth*. Cambridge, MA: Harvard University Press.

Gallaher, A. (1964) 'Plainville: the twice-studied town', in Vidich, A.J., Bensman, J. and Stein, M.R. (eds), *Reflections on Community Studies*. New York: John Wiley and Sons. pp. 285–303.

Gallo, A.M. and Knafl, K.A. (1998) 'Parents' reports of "tricks of the trade" for managing childhood chronic illness', *Journal of the Society of Paediatric Nurses*, 3 (3): 93–102.

Gates, M.F. and Hinds, P.S. (2000) 'Qualitative researchers working as teams', in Moch, S.D. and Gates, M.F. (eds), *The Researcher Experience in Qualitative Research*. Thousand Oaks: Sage. pp. 94–108.

Glaser, B.G. (1962) 'Secondary analysis: a strategy for the use of knowledge from research elsewhere', *Social Problems*, 10 (1): 70–74.

Glaser, B.G. (1963) 'Retreading research materials: the use of secondary analysis by the independent researcher', *The American Behavioural Scientist*, 6 (10): 11–14.

Glaser, B. and Strauss, A.L. (1967) *The Discovery of Grounded Theory: Strategies for Qualitative Research*. Chicago: Aldine Publishing Company.

Glass, G.V. (1976) 'Primary, secondary, and meta-analysis of research', *Educational Researcher*, 5 (11): 3–8.

Gleit, C. and Graham, B. (1989) 'Secondary data analysis: a valuable resource', *Nursing Research*, 38 (6): 380–81.

Goldthorpe, J.H. (1980) *Social Mobility and Class Structure in Modern Britain*. Oxford: Clarendon Press.

Goldthorpe, J.H., Lockwood, D., Bechhofer, F. and Platt, J. (1968a) *The Affluent Worker: Industrial Attitudes and Behaviour*. London: Cambridge University Press.

Goldthorpe, J.H., Lockwood, D., Bechhofer, F. and Platt, J. (1968b) *The Affluent Worker: Political Attitudes and Behaviour*. London: Cambridge University Press.

Goldthorpe, J.H., Lockwood, D., Bechhofer, F. and Platt, J. (1969) *The Affluent Worker in the Class Structure*. London: Cambridge University Press.

Gooding, B.A. (1988) 'Secondary analysis: a method for learning research activities', *Journal of Nursing Education*, 27 (5): 229–30.

Gregory, D. and Longman, A. (1992) 'Mothers' suffering: sons who died of AIDS', *Qualitative Health Research*, 2 (3): 334–57.

Groeschel, R. (2000) 'The interview archive "Youth in the 20th century" at the POSOPA e.V.', *Forum: Qualitative Social Research* [Online Journal], 1 (3): English abstract; full text available in German only. Available at http://qualitative-research.net/fqs/fqs-eng.htm [accessed 28/8/2003].

Guba, E.G. (1981) 'Criteria for assessing the trustworthiness of naturalistic inquiries', *Educational Communication & Technology Journal*, 29: 75–91.

Guba, E.G. and Lincoln, Y.S. (1989) *Fourth Generation Evaluation*. Thousand Oaks: Sage.

Hakim, C. (1982) *Secondary Analysis in Social Research: A Guide to Data Sources and Methods with Examples*. London: George Allen Unwin.

Hall, J.M., Stevens, P.E. and Meleis, A.I. (1992) 'Developing the construct of role integration: a narrative analysis of women clerical workers' daily lives', *Research in Nursing & Health*, 15: 447–57.

Halpern, E.S. (1983) 'Auditing naturalistic inquiries: the development and application of a model'. Unpublished thesis: Indiana University.

Hammersley, M. (1990) *Reading Ethnographic Research: A Critical Guide*. London: Longman. (First edition).

Hammersley, M. (1997a) 'Qualitative data archiving: some reflections on its prospects and problems', *Sociology*, 31 (1): 131–42.

Hammersley, M. (1997b) 'The relationship between qualitative and quantitative research: paradigm loyalty versus methodological eclecticism', in Richardson, J.T. (ed.) *Handbook of Qualitative Research Methods for Psychology and the Social Sciences*. Leicester: BPS Books.

Hammersley, M. (1999) 'Not bricolage but boatbuilding: exploring two metaphors for thinking about ethnography', *Journal of Contemporary Ethnography*, 28 (5): 574–85.

Hauser, R.M. (1987) 'Sharing data: it's time for ASA journals to follow the folkways of a scientific sociology', *American Sociological Review*, 52 (6): vi–viii.

Heaton, J. (1998) 'Secondary analysis of qualitative data', *Social Research Update*, 22.

Heaton, J. (2000) 'Secondary analysis of qualitative data: a review of the literature'. York: Social Policy Research Unit (SPRU), University of York.

Heaton, J. (2001) 'Hospital discharge and the temporal regulation of bodies', *Time & Society*, 10 (1): 93–111.

Heaton, J., Arksey, H. and Sloper, P. (1999) 'Carers' experiences of hospital discharge and continuing care in the community', *Health and Social Care in the Community*, 7 (2): 91–99.

Hedrick, T.E. (1988) 'Justifications for the sharing of social science data', *Law and Human Behaviour*, 12 (2): 163–71.

Herron, D.G. (1989) 'Secondary data analysis: research method for the clinical nurse specialist', *Clinical Nurse Specialist*, 3 (2): 66–69.

Hilgartner, S. and Brandt-Rauf, S.I. (1994) 'Data access, ownership, and control: toward empirical studies of access practices', *Knowledge: Creation, Diffusion, Utilization*, 15 (4): 355–72.

Hill, M.R. (1993) *Archival Strategies and Techniques*. Newbury Park: Sage.

Hinds, P.S., Vogel, R.J. and Clarke-Steffen, L. (1997) 'The possibilities and pitfalls of doing a secondary analysis of a qualitative data set', *Qualitative Health Research*, 7 (3): 408–24.

Hogue, C.J.R. (1991) 'Ethical issues in sharing epidemiologic data', *Journal of Clinical Epidemiology*, 44 (1): 103S-07S.

Holmes, L. (1957) 'A restudy of Manu'an culture: a problem in methodology'. Unpublished thesis: Northwestern University.

Hood-Williams, J. and Harrison, W.C. (1998) '"It's all in the small print ...": archiving and qualitative research', *Network – Newsletter of the British Sociological Association*, (70): 8–9.

Howarth, K. (1998) *Oral History: A Handbook*. Stroud: Sutton Publishing.

Humphrey, C.K., Estabrooks, C.A., Norris, J.R., Smith, J.E. and Hesketh, K.L. (2000) 'Archivist on board: contributions to the research team', *Forum: Qualitative Social*

Research [Online Journal], 1 (3): 19 paragraphs. Available at http://qualitative-research. net/fqs/fqs-eng.htm [accessed 28/8/2003].
Hutchby, I. and Woofit, R. (1998) *Conversation Analysis: Principles, Practices and Applications.* Cambridge: Polity Press.
Hutchinson, S. (1987) 'Self-care and job stress', *Image: Journal of Nursing Scholarship*, 19 (4): 192–96.
Hutchinson, S.A. (1990) 'Responsible subversion: a study of rule-bending among nurses', *Scholarly Inquiry for Nursing Practice: An International Journal*, 4 (1): 3–17.
Hyman, H.H. (1972) *Secondary Analysis of Sample Surveys: Principles, Procedures, and Potentialities.* New York: John Wiley & Sons, Inc.
Inter-university Consortium for Political and Social Research (2002) 'Guide to Social Science Data Preparation and Archiving.' Available at http://www.icpsr.umich. edu/ACCESS/dpm.htm [accessed 18/2/2003].
International Sociological Association (2001) 'Code of ethics.' Available at http://www.ucm.es/info/isa/about/isa_code_of_ethics.htm [accessed 10/2/2003].
International Tribunal of Nuremberg 1947 (1990) 'The Nuremberg Code', in Connor, S.S. and Fuenzalida-Puelma, H.L. (eds), *Bioethics: Issues and Perspectives.* Washington DC: Pan American Health Organisation. pp. 217–18 (reprinted in appendices).
Jackson, J.E. (1990) '"I am a fieldnote": fieldnotes as a symbol of professional identity', in Sanjek, R. (ed.) *Fieldnotes: The Makings of Anthropology.* Ithaca: Cornell University Press. pp. 3–33.
Jacobson, A.F., Hamilton, P. and Galloway, J. (1993) 'Obtaining and evaluating data sets for secondary analysis in nursing research', *Western Journal of Nursing Research*, 15 (4): 483–94.
Jairath, N. (1999) 'Myocardial infarction patients' use of metaphors to share meaning and communicate underlying frames of experience', *Journal of Advanced Nursing*, 29 (2): 283–89.
James, J.B. and Sørensen, A. (2000) 'Archiving longitudinal data for future research: why qualitative data add to a study's usefulness', *Forum: Qualitative Social Research* [Online Journal], 1 (3): 57 paragraphs. Available at http://qualitative-research. net/fqs/fqs-eng.htm [accessed 28/8/2003].
Janesick, V.J. (1994) 'The dance of qualitative research design: metaphor, methodolatory and meaning', in Denzin, N.K. and Lincoln, Y.S. (eds), *Handbook of Qualitative Research.* Thousand Oaks, CA: Sage. pp. 209–19.
Jenny, J. and Logan, J. (1996) 'Caring and comfort metaphors used by patients in critical care', *Image: Journal of Nursing Scholarship*, 28 (4): 349–52.
Jensen, L.A. and Allen, M.N. (1996) 'Meta-synthesis of qualitative findings', *Qualitative Health Research*, 6 (4): 553–60.
Jones, J.B. (1997) 'Representations of menopause and their health care implications: a qualitative study', *American Journal of Preventive Medicine*, 13 (1): 58–65.
Jones, C. and Rupp, S. (2000) 'Understanding the carers' world: a biographical-interpretive case study', in Chamberlayne, P., Bornat, J. and Wengraf, T. (eds), *The Turn to Biographical Methods in Social Science: Comparative Issues and Examples.* London: Routledge. pp. 276–89.
Jupp, V. and Norris, C. (1993) 'Traditions in documentary analysis', in Hammersley, M. (ed.) *Social Research: Philosophy, Politics and Practice.* London: Sage. pp. 37–51.

Katz, A. (1997) '"Mom, I have something to tell you" – disclosing HIV infection', *Journal of Advanced Nursing*, 25: 139–43.

Kearney, M.H., Murphy, S. and Rosenbaum, M. (1994a) 'Mothering on crack cocaine: a grounded theory analysis', *Social Science & Medicine*, 38 (2): 351–61.

Kearney, M.H., Murphy, S. and Rosenbaum, M. (1994b) 'Learning by losing: sex and fertility on crack cocaine', *Qualitative Health Research*, 4 (2): 142–62.

Kidd, P., Scharf, T. and Veazie, M. (1996) 'Linking stress and injury in the farming environment: a secondary analysis of qualitative data', *Health Education Quarterly*, 23 (2): 224–37.

Kiecolt, K.J. and Nathan, L.E. (1985) *Secondary Analysis of Survey Data*. London: Sage.

Kirk, J. and Miller, M. (1986) *Reliability and Validity in Qualitative Research*. Beverly Hills, CA: Sage.

Kirschbaum, M.S. and Knafl, K. (1996) 'Major themes in parent-provider relationships: a comparison of life-threatening and chronic illness experience', *Journal of Family Nursing*, 2 (2): 195–216.

Knafl, K.A., Ayres, L., Gallo, A.M., Zoeller, L.H. and Breitmayer, B.J. (1995) 'Learning from stories: parents' accounts of the pathways to diagnosis', *Pediatric Nursing*, 21 (5): 411–15.

Konopásek, Z. and Kusá, Z. (2000) 'Re-use of life stories in an ethnomethodological research', *Forum: Qualitative Social Research* [Online Journal], 1 (3): 42 paragraphs. Available at http://qualitative-research.net/fqs/fqs-eng.htm [accessed 28/8/2003].

Kuula, A. (2000) 'Making qualitative data fit the "Data Documentation Initiative" or vice versa?', *Forum: Qualitative Social Research* [Online Journal], 1 (3): 28 paragraphs. Available at http://qualitative-research.net/fqs/fqs-eng.htm [accessed 28/8/2003].

Langness, L. (1965) *The Life History in Anthropological Science*. New York: Holt, Rinehart and Winston.

Langness, L.L. and Frank, G. (1981) *Lives: An Anthropological Approach to Biography*. Novato, CA: Chandlet & Sharp.

Lapadat, J.C. (2000) 'Problematizing transcription: purpose, paradigm and quality', *International Journal Social Research Methodology*, 3 (3): 203–19.

LeCompte, M.D. and Goetz, J.P. (1982) 'Problems of reliability and validity in ethnographic research', *Review of Educational Research*, 52 (1): 31–60.

Leh, A. (2000) 'Problems of archiving oral history interviews. The example of the archive "German memory"', *Forum: Qualitative Social Research* [Online Journal], 1 (3): 26 paragraphs. Available at http://qualitative-research.net/fqs/fqs-eng.htm [accessed 28/8/2003].

Levi-Strauss, C. (1976) *The Savage Mind*. London: Weidenfeld and Nicolson. (First published in English 1966, in French 1962).

Lewis, O. (1951) *Life in a Mexican Village: Tepoztlan Restudied*. Urbana: The University of Illinois Press.

Lincoln, Y.S. (1995) 'Emerging criteria for quality in qualitative and interpretive research', *Qualitative Inquiry*, 1 (3): 275–89.

Lincoln, Y.S. and Guba, E.G. (1985) *Naturalistic Inquiry*. Newbury Park: Sage.

Lobo, M.L. (1986) 'Secondary analysis as a strategy for nursing research', in Chinn, P.L. (ed.) *Nursing Research. Issues and Implementation*. Rockville: Aspen Publishers. pp. 295–304.

Logan, J. and Jenny, J. (1990) 'Deriving a new nursing diagnosis through qualitative research: dysfunctional ventilatory weaning response', *Nursing Diagnosis*, 1 (1): 37–43.

Lynd, R.S. and Lynd, H.M. (1929) *Middletown: A Study in Contemporary American Culture*. London: Constable & Co Ltd.

Lynd, R.S. and Lynd, H.M. (1937) *Middletown in Transition: A Study in Cultural Conflicts*. New York: Harcourt, Brace and Company.

Marshall, C. and Rossman, G.B. (1999) *Designing Qualitative Research*. Thousand Oaks: Sage. (Third edition).

Marz, K. and Dunn, C.S. (2000) 'Depositing data with the Data Resources Program of the National Institute of Justice: a handbook'. Ann Arbor: Inter-university Consortium for Political and Social Research (ICPSR).

Mauthner, N.S., Parry, O. and Backett-Milburn, K. (1998) 'The data are out there, or are they? Implications for archiving and revisiting qualitative data', *Sociology*, 32 (4): 733–45.

May, T. (ed.) (2002) *Qualitative Research In Action*. London: Sage.

McArt, E.W. and McDougal, L.W. (1985) 'Secondary data analysis – a new approach to nursing research', *Image: The Journal of Nursing Scholarship*, XVII (2): 54–57.

McLaughlin, E. and Ritchie, J. (1994) 'Legacies of caring: the experiences and circumstances of ex-carers', *Health and Social Care in the Community*, 2: 241–53.

McLeod, J. (2000) 'Qualitative Research as bricolage', Paper presented at Society for Psychotherapy Research Annual Conference. Chicago 22 June. Available at http:shs.tay.ac.uk/shtjm/Qualitative%20Research%20as%20Bricolage.html [accessed 28/8/2003].

Mead, M. (1928) *Coming of Age in Samoa: A Psychological Study of Primitive Youth for Western Civilisation*. New York: William Morrow.

Medical Research Council (2000) 'Good research practice'. MRC Ethics Series. London: Medical Research Council.

Medical Research Council of Canada, Natural Sciences and Engineering Research Council of Canada and Social Sciences and Humanities Research Council of Canada (1998) 'Tri-council policy statement: ethical conduct for human research involving humans.' Available at http://www.nserc.ca/programs/ethics/english/english/policy.htm [accessed 10/2/2003].

Melton, G.B. (1988a) 'When scientists are adversaries, do participants lose?', *Law and Human Behaviour*, 12 (2): 191–98.

Melton, G.B. (1988b) 'Must researchers share their data?', *Law and Human Behaviour*, 12 (2): 159–62.

Messias, D.K.H., Yeager, K.A., Dibble, S.L. and Dodd, M.J. (1997) 'Patients' perspectives of fatigue while undergoing chemotherapy', *Oncology Nursing Forum*, 24 (1): 43–48.

Miles, M.B. and Huberman, A.M. (1994) *Qualitative Data Analysis: An Expanded Sourcebook*. Thousand Oaks: Sage. (Second edition).

Miller, D. (1967) 'Retrospective analysis of posthospital mental patients' worlds', *Journal of Health & Social Behaviour*, 8 (2): 136–40.

Miller, J.D. (1982) 'Secondary analysis and science education research', *Journal of Research in Science Teaching*, 19 (9): 719–25.

Miller, G. and Dingwall, R. (eds) (1997) *Context and Method in Qualitative Research*. London: Sage.

Miller, R.L. (2000) *Researching Life Stories and Family Histories*. London: Sage.

Muhr, T. (2000) 'Increasing the reusability of qualitative data with XML', *Forum: Qualitative Social Research* [Online Journal], 1 (3): 64 paragraphs. Available at http://qualitative-research.net/fqs/fqs-eng.htm [accessed 28/8/2003].

Murphy, E., Dingwall, R., Greatbatch, D., Parker, S. and Watson, P. (1998) 'Qualitative research methods in health technology assessment: a review of the literature', *Health Technology Assessment*, 2 (16).

National Children's Bureau (undated) 'Guidelines for research.' Available at http://www.ncb.org.uk/resguide.htm [accessed 20/12/2001].

National Institutes of Health (1995) 'Guidelines for the conduct of research involving human subjects at the National Institutes of Health.' Available at http://ohsr.od.nih.gov/guidelines.php3 [accessed 20/3/2002].

Newson, J. and Newson, E. (1963) *Infant Care in an Urban Community*. London: George Allen and Unwin.

Newson, J. and Newson, E. (1968) *Four Years Old in an Urban Community*. London: George Allen and Unwin.

Newson, J. and Newson, E. (1975) *Seven Years Old in the Home Environment*. London: George Allen and Unwin.

Noblit, G. and Hare, R. (1988) *Meta-ethnography: Synthesising Qualitative Studies*. Newbury Park, CA: Sage.

Office for Protection from Research Risks (undated) 'Institutional Review Board Guidebook.' Available at http://ohrp.osophs.dhhs.gov/irb_preface.htm [accessed 20/3/2002].

Oral History Society (undated) 'Ethical guidelines.' Available at http://www.oralhistory.org.uk/ethics/ethics.html [accessed 10/2/2003].

Ottenberg, S. (1990) 'Thirty years of fieldnotes: changing relationships to the text', in Sanjek, R. (ed.) *Fieldnotes: The Makings of Anthropology*. Ithaca: Cornell University Press. pp. 139–60.

Paget, M.A. (1983) 'On the work of talk: studies in misunderstandings', in Fisher, S. and Todd, A.D. (eds), *The Social Organisation of Doctor-Patient Communication*. Washington: The Center for Applied Linguistics. pp. 54–74.

Pascalev, A. (1996) 'Images of death and dying in the intensive care unit', *Journal of Medical Humanities*, 17 (4): 219–36.

Paterson, B.L., Thorne, S.E., Canam, C. and Jillings, C. (2001) *Meta-Study of Qualitative Health Research: A Practical Guide to Meta-Analysis and Meta-Synthesis*. Thousand Oaks: Sage.

Patton, M.Q. (2002) *Qualitative Research and Evaluation Methods*. Thousand Oaks: Sage. (Third edition).

Pickens, J.M. (1999) 'Living with serious mental illness: the desire for normalcy', *Nursing Science Quarterly*, 12 (3): 233–39.

Plummer, K. (1983) *Documents of Life: An Introduction to the Problems and Literature of a Humanistic Method*. London: George Allen & Unwin.

Plummer, K. (2001) *Documents of Life 2: An Invitation to a Critical Humanism*. London: Sage.

Popay, J., Rogers, A. and Williams, G. (1998) 'Rationale and standards for the systematic review of qualitative literature in health services research', *Qualitative Health Research*, 8 (3): 341–51.

Pope, L. (1942) *Millhands and Preachers: A Study of Gastonia*. New Haven: Yale University Press.

Pope, C. and Mays, N. (eds) (2000) *Qualitative Research in Health Care*. BMJ Publishing Group: London. (Second edition).

Pope, C., Ziebland, S. and Mays, N. (2000) 'Analysing qualitative data', *British Medical Journal*, 320: 114–16.

Popkess-Vawter, S., Brandau, C. and Straub, J. (1998) 'Triggers of overeating and related intervention strategies for women who weight cycle', *Applied Nursing Research*, 11 (2): 69–76.

Powers, B.A. (1996) 'Relationships among older women living in a nursing home', *Journal of Women & Aging*, 8 (3–4): 179–98.

Princeton, J.C. (1993) 'Education for executive nurse administrators: a databased curricular model for doctoral (PhD) programs', *Journal of Nursing Education*, 32 (2): 59–63.

Procter, M. (1995) 'Analysing other researchers' data', in Gilbert, N. (ed.) *Researching Social Life*. London: Sage. pp. 255–69.

Public Records Office (2000) 'Data Protection Act 1998. A guide for records managers and archivists.' Available at http://www.pro.gov.uk/recordsmanagement/dp/dpguide.pdf [accessed 20/3/2002].

Qualidata (undated a) 'Legal and ethical issues in interviewing.' Available at http://www.qualidata.essex.ac.uk/creatingData/guidelines.asp [accessed 20/11/2002].

Qualidata (undated b) 'Legal and ethical issues in interviewing children.' Available at http://www.qualidata.essex.ac.uk/creatingData/guidelineschildren.asp [accessed 20/11/2002].

Qualidata (undated c) 'Legal issues.' Available at http://www.qualidata.essex.ac.uk/creatingData/legal.asp#copyright [accessed 18/2/2003].

Ramcharan, P. and Cutcliffe, J.R. (2001) 'Judging the ethics of qualitative research: considering the "ethics as process" model', *Health and Social Care in the Community*, 9 (6): 358–66.

Redfield, R. (1930) *Tepoztlan – A Mexican Village: A Study of Folk Life*. Chicago: University of Chicago Press.

Reed, J. (1992) 'Secondary data in nursing research', *Journal of Advanced Nursing*, 17: 877–83.

Rehm, R.S. and Catanzaro, M.L. (1998) '"It's just a fact of life": family members' perceptions of parental chronic illness', *Journal of Family Nursing*, 4 (1): 21–40.

Reid, A. and Gough, S. (2000) 'Guidelines for reporting and evaluating qualitative research: what are the alternatives?', *Environmental Education Research*, 6 (1): 59–91.

Reynolds, P.D. (1982) *Ethics and Social Research*. New Jersey: Prentice-Hall.

Richardson, L. (1992) 'Trash on the corner: ethics and technography', *Journal of Contemporary Ethnography*, 21 (1): 103–19.

Richardson, J.T. (ed.) (1996) *Handbook of Qualitative Research Methods for Psychology and the Social Sciences*. Leicester: BPS Books.

Riedel, M. (2000) *Research Strategies for Secondary Data: A Perspective for Criminology and Criminal Justice*. Thousand Oaks: Sage.

Ritchie, J. and Spencer, L. (1995) 'Qualitative data analysis for applied policy research', in Bryman, A. and Burgess, R. (eds), *Analysing Qualitative Data*. London: Routledge. pp. 173–94.

Rivard, J.C., Johnsen, M.C., Morrissey, J.P. and Starrett, B.E. (1999) 'The dynamics of interagency collaboration: how linkages develop for child welfare and juvenile justice sectors in a system of care demonstration', *Journal of Social Service Research*, 25 (3): 61–82.

Robinson, C.A. (1993) 'Managing life with a chronic condition: the story of normalisation', *Qualitative Health Research*, 3 (1): 6–28.
Robinson, C.A. and Thorne, S. (1984) 'Strengthening family "interference"', *Journal of Advanced Nursing*, 9: 597–602.
Rolph, S. (2000) 'Legal and ethical issues in interviewing people with learning difficulties.' Available at http://www.qualidata.essex.ac.uk/creatingData/guidelines learningdifficulty.asp [accessed 20/11/2002].
Rose, H. (2001) 'The commodification of bioinformation. the Icelandic Health Sector Database'. London: The Wellcome Trust.
Rush, K.L. and Ouellet, L.L. (1997) 'Mobility aids and the elderly client', *Journal of Gerontological Nursing*, 23 (1): 7–15.
Sandelowski, M. (1994a) 'Separate, but less unequal: fetal ultrasonography and the transformation of expectant mother/fathehood', *Gender & Society*, 8 (2): 230–45.
Sandelowski, M. (1994b) 'Channel of desire: fetal ultrasonograpy in two use-contexts', *Qualitative Health Research*, 4 (3): 262–80.
Sandelowski, M. (1997) '"To be of use": enhancing the utility of qualitative research', *Nursing Outlook*, 45: 125–32.
Sandelowski, M. and Black, B.P. (1994) 'The epistemology of expectant parenthood', *Western Journal of Nursing Research*, 16 (6): 601–22.
Sandelowski, M. and Jones, L.C. (1996) 'Couples' evaluations of foreknowledge of fetal impairment', *Clinical Nursing Research*, 5 (1): 81–96.
Sanjek, R. (ed.) (1990a) *Fieldnotes: The Makings of Anthropology*. Ithaca: Cornell University Press.
Sanjek, R. (1990b) 'Fieldnotes and others', in Sanjek, R. (ed.) *Fieldnotes: The Makings of Anthropology*. Ithaca: Cornell University Press. pp. 324–40.
Scheff, T.J. (1986) 'Toward resolving the controversy over "thick description"', *Current Anthropology*, 27 (4): 408–9.
Schwandt, T.A. and Halpern, E.S. (1988) *Linking Auditing and Metaevaluation: Enhancing Quality in Applied Research*. Newbury Park: Sage.
Scott, J. (1990) *A Matter of Record: Documentary Sources in Social Research*. Cambridge: Polity Press.
Seale, C. (1999) *The Quality of Qualitative Research*. London: Sage.
Sheridan, D. (2000) 'Reviewing Mass-Observation: the archive and its researchers thirty years on', *Forum: Qualitative Social Research* [Online Journal], 1 (3): 9 paragraphs. Available at http://qualitative-research.net/fqs/fqs-eng.htm [accessed 28/8/2003].
Sieber, J.E. (1988) 'Data sharing: defining problems and seeking solutions', *Law and Human Behaviour*, 12 (2): 199–206.
Sieber, J.E. (ed.) (1991) *Sharing Social Science Data: Advantages and Challenges*. London: Sage.
Silverman, D. (1993) *Interpreting Qualitative Data: Methods for Analysing Talk, Text and Interaction*. London: Sage.
Silverman, D. (2000) *Doing Qualitative Research: A Practical Handbook*. London: Sage.
Smith, G.D. and Egger, M. (1998) 'Meta-analysis: unresolved issues and future developments', *British Medical Journal*, 316 (7126): 221–25.
Smith, J.K. (1984) 'The problem of criteria for judging interpretive inquiry', *Educational Evaluation and Policy Analysis*, 6 (4): 379–91.
Social Research Association (2002) 'Ethical Guidelines.' Available at http://www.the-sra.org.uk/index2.htm [accessed 10/2/2003].

Society for Research in Child Development (1990–91) 'Ethical standards for research with children.' Available at http://www.srcd.org/about.html [accessed 15/2/2003].

Spencer, H. (1873–1881) *Descriptive Sociology; or, Groups of Sociological Facts*. London: Williams and Norgate. (Compiled and abstracted by Duncan, D., Scheppig, R. and Collier, J.).

Stacey, M. (1960) *Tradition and Change: A Study of Banbury*. London: Oxford University Press.

Stacey, M., Batstone, E. and Bell, C. (1975) *Power, Persistence and Change: A Second Study of Banbury*. London: Routledge and Kegan Paul.

Stanley, B. and Stanley, M. (1988) 'Data sharing: the primary researcher's perspective', *Law and Human Behaviour*, 12 (2): 173–80.

Sterling, T.D. (1988) 'Analysis and reanalysis of shared scientific data', *Annals of the American Academy of Political & Social Science*, 495: 49–60.

Stewart, D.W. and Kamins, M.A. (1993) *Secondary Research: Information Sources and Methods*. Newbury Park: Sage. (Second edition).

Strauss, A. and Corbin, J. (1990) *Basics of Qualitative Research: Grounded Theory Procedures and Techniques*. Newbury Park: Sage.

Strauss, A. and Corbin, J. (1998) *Basics of Qualitative Research: Techniques and Procedures For Developing Grounded Theory*. Thousand Oaks, CA: Sage.

Strauss, A.L. and Glaser, B.G. (1977) *Anguish: A Case History of a Dying Trajectory*. London: Martin Robertson.

Szabo, V. and Strang, V.R. (1997) 'Secondary analysis of qualitative data', *Advances in Nursing Science*, 20 (2): 66–74.

Szabo, V. and Strang, V.R. (1999) 'Experiencing control in caregiving', *Image: Journal of Nursing Scholarship*, 31 (1): 71–75.

Temple, B. (1998) 'A fair trial? Judging quality in qualitative research.' *International Journal of Social Research Methodology*, 1 (3): 205–15.

ten Have, P. (1999) *Doing Conversation Analysis: A Practical Guide*. London: Sage.

The National Committee for Research Ethics in the Social Sciences and the Humanities – The Research Council of Norway (NESH) (2001) 'Guidelines for research ethics in the social sciences, law and humanities.' Available at http:www.etikkom.no/NESH/guidelines.htm [accessed 20/3/2002].

The Royal Liverpool Children's Inquiry: report (2001) London: The Stationary Office. Available at: http:www.rlcinquiry.org.uk [accessed 28/8/2003].

The Wellcome Trust (2002) 'Guidelines on Good Research Practice.' Available at http://www.wellcome.ac.uk/en/images/GOODRESEARCH_5815_5900.pdf [accessed 5/9/2002].

Thomas, W. and Znaniecki, F. (1958) *The Polish Peasant in Europe and America*. New York: Dover Publications.

Thompson, P. (1998) 'Sharing and reshaping life stories: problems and potential in archiving research narratives', in Chamberlain, M. and Thompson, P. (eds), *Narrative and Genre*. London: Routledge. pp. 167–81.

Thompson, P. (2000a) 'Re-using qualitative research data: a personal account', *Forum: Qualitative Social Research* [Online Journal], 1 (3): 48 paragraphs. Available at http://qualitative-research.net/fqs/fqs-eng.htm [accessed 28/8/2003].

Thompson, P. (2000b) *The Voice of the Past: Oral History*. Oxford: Oxford University Press. (Third Edition).

Thorne, S.E. (1988) 'Helpful and unhelpful communications in cancer care: the patient perspective', *Oncology Nursing Forum*, 15 (2): 167–72.

Thorne, S.E. (1990a) 'Navigating troubled waters: chronic illness experience in a health care crisis', unpublished thesis, The Union Institute of Advanced Studies: Cincinnati.
Thorne, S.E. (1990b) 'Constructive noncompliance in chronic illness', *Holistic Nursing Practice*, 5 (1): 62–69.
Thorne, S.E. (1990c) 'Mothers with chronic illness: a predicament of social construction', *Health Care for Women International*, 11: 209–21.
Thorne, S.E. (1994) 'Secondary analysis in qualitative research: issues and implications', in Morse, J.M. (ed.) *Critical Issues in Qualitative Research Methods*. London: Sage. pp. 263–79.
Thorne, S. (1998) 'Ethical and representational issues in qualitative secondary analysis', *Qualitative Health Research*, 8 (4): 547–55.
Thorne, S.E., Harris, S.R., Hislop, T.G. and Vestrup, J.A. (1999) 'The experience of waiting for diagnosis after an abnormal mammogram', *The Breast Journal*, 5 (1): 42–51.
Thorne, S.E., Nyhlin, K.T. and Paterson, B.L. (2000) 'Attitudes toward patient expertise in chronic illness', *International Journal of Nursing Studies*, 37 (4): 303–11.
Thorne, S.E. and Robinson, C.A. (1988a) 'Health care relationships: the chronic illness perspective', *Research in Nursing & Health*, 11: 293–300.
Tishelman, C. and Sachs, L. (1998) 'The diagnostic process and the boundaries of normality', *Qualitative Health Research*, 8 (1): 48–60.
Tizard, B. and Hughes, M. (1984) *Young Children Learning*. London: Fontana.
Turner, J.H. (1985) *Herbert Spencer: A Renewed Appreciation*. Beverly Hills: Sage.
UK Data Archive (undated) 'Qualidata process guide: qualitative data processing.' Available at http://www.qualidata.essex.ac.uk/about/document.asp [accessed 18/2/2003].
Vallerand, A.H. and Ferrell, B.R. (1995) 'Issues of control in patients with cancer pain', *Western Journal of Nursing Research*, 17 (5): 467–83.
Vidich, A.J. (1992) 'Boston's North End: an American epic', *Journal of Contemporary Ethnography*, 21 (1): 80–102.
Vidich, A.J. and Lyman, S.M. (1994) 'Qualitative methods: their history in sociology and anthropology', in Denzin, N.K. and Lincoln, Y.S. (eds), *Handbook of Qualitative Research*. Thousand Oaks: Sage. pp. 23–59.
Walkerdine, V. and Lucey, H. (1989) *Democracy in the Kitchen: Regulating Mothers and Socialising Daughters*. London: Virago Press.
Ward, A. (undated) 'Copyright and oral history.' Available at http://www.oralhistory.org.uk/ethics/ [accessed 10/2/2003].
Wax, M.L. (1997) 'On negating positivism: an anthropological dialectic', *American Anthropologist*, 99 (1): 17–23.
Weaver, A. (1994) 'Deconstructing dirt and disease: the case of TB', in Bloor, M. and Taraborrell, P. (eds), *Qualitative Studies in Health and Medicine*. Aldershot: Avebury. pp. 76–95.
Weaver, A. and Atkinson, P. (1994) *Microcomputing and Qualitative Data Analysis*. Aldershot: Avebury.
West, J. (1945) *Plainville, USA*. New York: Columbia University Press.
West, J. and Oldfather, P. (1995) 'Pooled case comparison: an innovation for cross-case study', *Qualitative Inquiry*, 1 (4): 452–64.
Whyte, W.F. (1992) 'In defence of Street Corner Society', *Journal of Contemporary Ethnography*, 21 (1): 52–68.

Whyte, W.F. (1993) *Street Corner Society: The Social Structure of an Italian Slum*. Chicago: The University of Chicago Press. (Fourth edition. First published 1943).

Whyte, W.F. (1997) *Creative Problem Solving in the Field: Reflections on a Career*. Walnut Creek: AltaMira Press.

Willison, D.J., Keshavjee, K., Nair, K., Goldsmith, C. and Holbrook, A.M. (2003) 'Patient consent preferences for research uses of information in electronic medical records: interview and survey data', *British Medical Journal*, 326 (7385): 373–76.

Woods, N.F. (1988) 'Using existing data sources: primary and secondary analysis', in Woods, N.F. and Catanzaro, M. (eds), *Nursing Research: Theory and Practice*. St. Louis: The C.V. Mosby Company. pp. 334–52.

World Medical Association (1964) 'The Declaration of Helsinki: Recommendations Guiding Medical Doctors in Biomedical Research Involving Human Subjects.' Available at http://www.wma.net/e/policy/17–c_e.html [accessed 15/2/2003].

Yamashita, M. and Forsyth, D.M. (1998) 'Family coping with mental illness: an aggregate from two studies, Canada and United States', *Journal of the American Psychiatric Association*, 4 (1): 1–8.

Yow, V.R. (1994) *Recording Oral History: A Practical Guide for Social Scientists*. Thousand Oaks: Sage.

Zeitlyn, D. (2000) 'Archiving anthropology', *Forum: Qualitative Social Research* [Online Journal], 1 (3): 20 paragraphs. Available at http://qualitative-research.net/fqs/fqs-eng.htm [accessed 28/8/2003].

Zussman, R. (1992) *Intensive Care: Medical Ethics and the Medical Profession*. Chicago: The University of Chicago Press.

Key Names and Titles Index

Abel, E. 26
Adams, A. 16
Affluent Worker (Goldthorpe et al., 1968a, 1968b, 1969) 46
Ahmad, N. 107, 120
Aita, V.A. 41, 51
Alderson, P. 2, 26, 79, 80, 88, 112, 114
Allen, M.N. 10, 18
Allen and Overy 75, 76, 82, 84, 85
American Sociological Association (ASA) 25, 74–5, 77, 81, 87
Angell, R. 17
Angst, D.B. 48, 99
Anguish: A Case History of a Dying Trajectory (Strauss and Glaser, 1977) 18
Arksey, H. 49, 107
ASA *see* American Sociological Association
Ashmore, M. 67
Association of Social Anthropologists of the UK and the Commonwealth 25, 74–5, 78, 81–2, 83–4
Atkinson, P. 41, 61–2, 64, 99, 105, 114, 115
Atkinson, R. 6, 7, 17, 21, 33
Austin, E.W. 26–7, 29
Australian National Health and Medical Research Council 75

Backett, K.C. 48
Backhouse, G. viii, 79, 83, 88
Baron, J.N. 26, 28, 33
Baxter, M. 108
Beck, C. 100
Bevan, M. 41
Biklen, S.K. 54, 55, 108
Black, B.P. 44, 97
Bloor, M. 39–40, 41, 47, 48, 72, 101, 107
Boddy, M. 32, 53
Boelen, W.A.M. 46–7
Bogdan, R.C. 54, 55, 108

Bond, G.C. 63
Boruch, R. 26, 75
Brandt-Rauf, S.I. 26
Breckenridge, D.M. 45, 97
British Sociological Association (BSA) 25, 74–5, 78, 82, 87, 108
British Sociological Association Medical Sociology Group 103–4, 121, 131–4
Brooks, I. 39
Brown, R.B. 39
Bryman, A. 18, 55, 57, 60, 71, 112, 115
BSA *see* British Sociological Association
Bull, M.J. 49
Bulmer, M. 77
Burgess, R.G. 46
Burns, N. 101
Burstein, L. 16

Calder, A. 7, 21
Campbell, D. 100
Canadian Sociology and Anthropology Association 74
Cantanzaro, M.L. 42
Caplow, T. 46
Ceci, S.J. 26
Cecil, J.S. 26
Cherlin, A. 26
Clarke-Steffen, L. 42, 63, 81, 96, 101, 107
Clayton, D.L.K. 40, 98, 99, 107
Clifford, J. 113
Clubb, J.M. 26, 28
Cohen, M.H. 49, 51, 53, 68, 70, 101
Cohen, S. 107
Colby, A. 26
Committee on National Statistics 22–4
Connor, S.S. 86
Corbin, J. 40, 97
Cordray, D.S. 75
Corti, L. viii, 2, 10, 14, 17, 19, 21, 22, 26, 29, 30, 31, 32, 33, 61, 63, 76, 79, 83, 107, 111, 114, 120, 124

Crabtree, B. 41, 51
Creswell, J.W. 99
Cutcliffe, J.R. 87

Dale, A. viii, 1, 2, 4, 5, 9, 13, 17, 26, 60–1
Davies, C.A. 104, 108
Davison, C. 48
Deatrick, J.A. 43, 48, 99
Denzin, N.K. x, 16, 17, 47, 54, 57, 99, 101, 109, 112–14, 115, 116, 119, 123, 124
Descriptive Sociology (Spencer, 1873) 20
Devine, F. 46
Dingwall, R. 108
Dixon-Woods, M. 18
Doherty, W.J. 50, 53
Dunn, C.S. 25, 26–7, 29, 77
Durkheim, E. 3

Earle, J.R. 46
Economic and Social Research Council (ESRC) 75, 82, 84, 85
Egger, M. 11
Eisner, E.W. 70
Elder, G.H. 4, 17
Elliot, R. 104, 134
Erickson, K. 72
ESRC *see* Economic and Social Research Council
Estabrooks, C.A. 11, 12, 33, 48, 103

Fendrich, M. 41, 49, 59, 98
Ferrell, B.R. 43, 49, 59, 97, 98
Fielding, J.L. 28, 61, 63, 65, 99, 103, 107, 111
Fielding, N.G. 22, 28, 61, 63, 65, 92, 99, 103, 107, 111
Fienberg, S.E. 1, 22, 23, 24, 26, 32, 33, 94
Fink, A.S. 2, 19, 21, 26, 63, 111, 114
Fischer, C. 134
Fiske, D. 100
Forsyth, D.M. 48, 103, 107
Frank, G. 33
Freeman, D. 46
Fuenzalida-Puelma, H.L. 86

Gallaher, A. 46
Gallo, A.M. 43, 59, 98, 99
Gates, M.F. 72
Glaser, B.G. viii, 8, 18, 19, 58, 114
Glass, G.V. 9, 10, 16
Gleit, C. 16

Goetz, J.P. 99
Goldthorpe, J.H. 4, 46
Gooding, B.A. 16
Gough, S. 99, 100, 108
Graham, B. 16
Gregory, D. 53, 59, 97, 98
Groeschel, R. 21
Grove, S. 101
Guba, E.G. 56, 66, 68–70, 100, 101, 102

Hakim, C. 1, 4, 5, 8, 17, 26
Hall, J.M. 43, 97, 98
Halpern, E.S. 68, 69, 102
Hammersley, M. 2, 10, 18, 19, 26, 28, 30, 55, 57, 61, 69, 99, 112, 114, 115, 116–17, 119, 124
Handbook of Qualitative Research (Denzin and Lincoln, 1994) 112, 116
Hare, R. 11
Harrison, T. 21
Harrison, W.C. 2, 26, 79, 86
Hauser, R.M. 26, 32
Heaton, J. viii, ix, 1, 6, 11, 13, 19, 26, 30, 35, 39, 42, 53, 63, 64, 99, 103, 111, 114, 126, 127, 129
Hedrick, T.E. 1, 26
Heritage, J. 7
Herron, D.G. 3, 16, 18
Hilgartner, S. 26
Hill, M.R. 33
Hinds, P.S. viii, 2, 9, 19, 26, 30, 37, 42, 50, 58, 61, 63, 72, 87, 92–3, 95, 96, 105, 111, 114, 121
Hogue, C.J.R. 26
Holmes, L. 46
Hood-Williams, J. 2, 79, 86
Howarth, K. 33
Huberman, A.M. 57, 58, 60, 66, 72, 101, 108
Hughes, M. 46
Humphrey, C.K. 19, 95
Hutchby, J. 7, 67
Hutchinson, S.A. 50, 53, 59–60, 96, 98, 101
Hyman, Herbert H. ix, 1, 3, 8, 9, 13, 17, 18, 26, 53, 123

ICPSR *see* Inter-university Consortium for Political and Social Research
International Sociological Association 74
International Tribunal of Nuremberg 74

Inter-university Consortium for Political and Social Research (ICPSR) 25, 76, 88

Jackson, J.E. 33, 114
Jacobson, A.F. 3, 16
Jairath, N. 40–1, 59, 98
James, J.B. 2, 14, 19, 21, 34, 88, 114
Janesick, V.J. 101
Jenny, J. 41, 43, 59, 69, 96, 97, 98, 102
Jensen, L.A. 10, 18
Jones, C. 41
Jones, J.B. 45, 97
Jones, L.C. 51
Jupp, V. 6, 54

Kamins, M.A. 1
Kane, R.L. 49
Katz, A. 44
Kearney, M.H. 42, 44, 59, 97–8
Kidd, P. 70, 101
Kiecolt, K.J. 1, 10, 13, 26
Kirk, J. 99
Kirschbaum, M.S. 48
Knafl, K. 42, 43, 48, 59, 97, 98, 99
Konopásek, Z. 33
Kusá, Z. 33
Kuula, A. 19, 21

Langness, L. 7, 33
Lapadat, J.C. 67
LeCompte, M.D. 99
Leh, A. 19, 21, 29, 83, 91, 96, 114
Levi-Strauss, C. 116, 124
Lewis, O. 30, 46
Life in a Mexican Village: Tepoztlan Restudied (Lewis, 1951) 30, 46
Lincoln, Y.S. x, 54, 56, 57, 66, 68–70, 99, 100, 101, 102, 109, 112–14, 115, 116, 119, 123, 124, 134
Lobo, M.L. 3, 17
Logan, J. 41, 43, 59, 69, 96, 97, 98, 102
Longman, A. 53, 59, 97, 98
Lucey, H. 46
Lyman, S.M. 32, 112, 115
Lynd, H.M. 46
Lynd, R.S. 46

Marcus, G. 113
Margaret Mead and Samoa (Freeman, 1983) 46
Marshall, C. 69

Marz, K. 25, 77
Mauthner, N.S. 2, 14, 19, 26, 29, 31, 64–5, 105–6, 112, 114, 115, 121
May, T. 104
Mays, N. 99
McArt, E.W. 9, 13, 16, 30
McDougal, L.W. 9, 13, 16, 30
McIntosh, J. 39, 40, 47, 48, 107
McLaughlin, E. 42, 59, 98
McLeod, J. 116–17, 119, 124
Mead, M. 46
Medical Research Council of Canada 75
Medical Research Council (UK) 32, 75
Melton, G.B. 26
Messias, D.K.H. 44, 97, 103
Middletown (Lynd and Lynd, 1929) 46
Middletown in Transition (Lynd and Lynd, 1937) 46
Miles, M.B. 57, 58, 60, 66, 72, 101, 108
Miller, D. 3
Miller, G. 108
Miller, J.D. 13, 16
Miller, M. 99
Miller, R.L. 7, 33, 57
Muhr, T. 22
Murdock, G.P. 20
Murphy, E. 55, 99, 101, 104, 108, 112

NAS *see* National Academy of Sciences
Nathan, L.E. 1, 10, 13, 26
National Academy of Sciences (NAS) 22–4
National Children's Bureau, The 75, 87
National Institutes of Health 75
Newson, E. 5
Newson, J. 5
Noblit, G. 11
Norris, C. 6, 54

Office for National Statistics (ONS) 4–5
Office for Protection from Research Risks 87
Oldfather, P. 14, 48, 63–4, 98
ONS *see* Office for National Statistics
Oral History Society 74, 83
Ottenberg, S. 62
Ouellet, L.L. 45

Paget, M.A. 72
Pascalev, A. 41, 68
Paterson, B.L. 18
Patton, M.Q. 99, 108

Phelps, E. 26
Pickens, J.M. 49, 97
Plainville Fifteen Years Later (Gallaher, 1964) 46
Plainville, USA (West, 1945) 46
Plummer, K. 6, 17
The Polish Peasant in Europe and America (Thomas and Znaniecki, 1958) 17–18
Popay, J. 18, 104
Pope, C. 72, 99
Pope, L. 46
Popkess-Vawter, S. 45–6, 101
Powers, B.A. 42
Princeton, J.C. 43, 97
Procter, M. 4
Public Records Office 88

Qualidata 76, 79–80, 82, 83, 121

Ramcharan, P. 87
Redfield, R. 30, 46
Reed, D. 67
Reed, J. 3
Rehm, R.S. 42
Reid, A. 99, 100, 108
Rennie, D. 134
Research Council of Norway, The 75
Reynolds, P.D. 87
Richardson, J.T. 108
Richardson, L. 47
Riedel, M. 16, 26
Ritchie, J. 42, 58, 59, 69, 98
Rivard, J.C. 53
Robinson, C.A. 49–50, 97, 107
Rolph, S. 25, 76, 79–80, 121
Romyn, D.M. 33
Rose, H. 34
Rossman, G.B. 69
Rupp, S. 41
Rush, K.L. 45

Sachs, L. 53, 101
Sacks, H. 30, 67
Sandelowski, M. 28, 44, 51, 53, 97, 99, 114
Sanjek, R. 6, 62
Scheff, T.J. 30
Schwandt, T.A. 69
Scott, J. 6
Seale, C. 18, 69, 104, 123
Secondary Analysis of Sample Surveys: Principles, Procedures and Potentialities (Hyman, 1972) 1

Sheridan, D. 7, 21, 114
Sherman, J.J. 26
Sieber, J.E. 1, 23, 26
Silverman, D. 108
Sloper, P. 49, 107
Smith, G.D. 11
Smith, J.K. 99, 100
Social Research Association 33, 74–5, 77–8
Society for Research in Child Development 74
Sørensen, A. 2, 14, 19, 21, 34, 88, 114
Spencer, H. 20
Spencer, L. 58, 69
Stacey, M. 46
Stanley, B. 1, 26, 28
Stanley, M. 1, 26, 28
Sterling, T.D. 26
Stewart, D.W. 1
Stirling, P. 22, 92
Strang, V.R. viii, 2, 10, 14, 19, 26, 43, 63, 69, 81, 94, 97, 99, 101, 102, 105, 108, 111, 114, 121, 124
Strauss, A. viii, 18, 40, 58, 97
Street Corner Society (Whyte, 1993) 46–7
Stull, D. 72
Suicide: A Study in Sociology (Durkheim, 1952) 3
Szabo, V. viii, 2, 10, 14, 19, 26, 42, 63, 69, 81, 94, 97, 99, 101, 102, 105, 108, 111, 114, 121, 124

Taylor, L. 107
Temple, B. 99
ten Have, P. 7
Terman, L. 4
The Polish Peasant in Europe and America (Thomas and Znaniecki, 1958) 17–18
Thomas, W. 17
Thompson, P. viii, 2, 6, 14, 17, 19, 33, 61, 63, 88, 91, 107, 111, 114, 120, 124
Thorne, S.E. viii, 2, 11, 14, 19, 26, 37, 39, 42, 44, 48, 49–50, 51, 58, 61, 68, 69, 97, 100, 101, 102–3, 107, 108, 111, 114, 125, 126
Tishelman, C. 53, 101
Tizard, B. 46
Turner, J.H. 20

UK Data Archive 76, 121

Key Names and Titles Index

Vallerand, A.H. 43
Vidich, A.J. 32, 47, 112, 115

Walker, E. 26
Walkerdine, V. 46
Ward, A. 83, 88
Wax, M.L. 20, 115
Weaver, A. 39, 41, 61–2, 99
Wellcome Trust, The 32, 75
West, J. 14, 46, 48, 63–4, 98
Whyte, W.F. 46–7, 57
Willison, D.J. 88

Woods, N.F. 3, 8, 9, 10, 16, 18
Woofit, R. 7, 67
World Medical Association 74
Writing Culture (Clifford and Marcus, 1986) 113

Yamashita, M. 48, 103, 107
Yow, V.R. 7

Zeitlyn, D. 22, 92
Znaniecki, F. 17
Zussman, R. 41

Subject Index

access to data 26–7, 92–4
additional or new questions 8–9, 16, 90, 110–11, 118
adequacy, referential 66, 69–70, 111, 115, 122
administrative data 3
age of data set(s) 96
aggregated analysis 10–12, 48
amplified analysis 38, 39–40, 47–50, 52, 92, 99, 111, 117–18, 129, 130
amplified sampling 14, 48
analysis
  evaluation criterion 132–3
  modi operandi 96–103
analytic expansion 14, 42, 44
anonymization 29, 82–3
anti-foundationalism 100
archives
  accessibility 94
  data 20–2, 29, 70, 83, 91–2, 96
  developments 19, 20–2, 119–20
  electronic catalogues 91
  ethical and legal frameworks 74, 76–7
  re-using 14, 96
armchair induction 14
artefactual data *see* non-naturalistic data
assessment
  of re-usability 92–6
  of research quality 121
assorted analysis 38, 50–1, 52, 59, 90, 96, 111, 117–18, 129, 130
audit trails 66, 68–70, 71, 100, 102, 111, 115, 122
augmentation of data 59–60
authors 36, 37, 49, 104
auto-data 13–14, 36–7, 63–5, 120–1, 130
autobiographies 37, 51, 68, 70, 110, 117

blurred genres moment 112–13, 116
boat builder figure 117

bricolage 117, 118, 124
bricoleurs 109, 116–19

CAQ-DAS (computer-assisted qualitative data analysis software) 22, 99
case studies 116
census data 3, 5, 21
CESSDA *see* Council of European Social Science Data Archives
codes of practice 25, 74–5
commercial concerns 28
completeness of data sets 94–5
computing
  developments 19, 20–2
  qualitative data analysis 22, 66, 99
  see also electronic...; on-line...
conduct of literature review 127–30
confessional accounts 104
confidentiality
  agreements 73
  data sharing and retention 25
  ethics committees 76
  guidelines 75, 81–2
  issues 81–3
  narrative interviews 27
  see also informed consent
contemporary landscape comparison 116–19
context 30–1, 60–5, 123–4
  see also not having been there
conversation analysis 5, 7–8, 12–13, 15, 30, 65–7, 110, 111, 114–15, 122
copyright 75, 83–5
cost-effectiveness of data sharing 29
Council of European Social Science Data Archives (CESSDA) 32
crisis of legitimation 113
crisis of representation 113
cross-validation 14

Subject Index

Danish Data Archives (DDA) 21
database of patients' experiences project (DIPEx) 92
data donors 12, 25–6, 77, 94, 121
data fit 30, 57–60, 95–6, 111–12, 115
data protection issues 82, 85
data sets 1, 91–6, 104, 117–18, 129
data sharing
  conversation analysis 12–13
  electronic 92–3
  ethical standards 75
  formal 12, 23
  informal 12, 23–4, 93–4, 96, 120–1
  material resources 119–20
  problems and benefits 26–31
  promotion 22–6
  Qualidata 114, 120
  retention 22–6
  statistical 1, 22–6
DDA *see* Danish Data Archives
de-identification 82
  see also anonymization
Declaration of Helsinki 74
defamation 85
demonstration studies 41
design
  primary research 95
  secondary analysis 90–1
diaries 6, 63
digitization 22, 92, 120
DIPEx *see* database of patients' experiences project
documentary analysis 5, 6, 15, 110, 122

Economic and Social Data Service (ESDS) 32
electronic archive catalogues 91
electronic data sharing 22, 92–3
electronic records 123
engineer figure 116–17, 119
epistemological issues
  data fit 57–60
  fieldnotes 62
  intellectual resources 122–3
  nature of qualitative research 54–7
  not having been there 31, 60–2, 64–5, 102–3, 111–12, 115
  raw data publication 66–8
  re-using data 30, 61–5, 114–15
  verification 65–71
epistemological position 55, 57, 65, 111–12, 115

ESDS *see* Economic and Social Data Service
ethical and legal issues 25, 73–88, 103–4, 111–12
ethics committees 76
ethics evaluation criteria 133–4
ethnographic studies 11, 41, 49, 116
evaluation criteria, qualitative research papers 103–4, 131–4
exploratory studies 58

face to face interviews 37, 117
fieldnotes 22, 39, 47, 61–3, 92
finding relevant data sets 91–2
formal data sharing 12, 23
framework approach 58, 69
frameworks, legal and ethical 74–7
functions
  appraisal 129
  secondary analysis 8–12, 16
funding 23, 74
future moment 113
future of secondary analysis 109–25

General Household Survey (GHS) 4
grounded theory 50, 58, 97–8, 106, 115–16
guidelines 25, 74–6, 81–2, 103–4, 121–2

headnotes 62
historical data 51, 90, 91
Human Genome Project 28
Human Relations Area Files (HRAF) 20, 115
humanism 62–3

individual meta-analysis 11
induction 14
informal data sharing 12, 23, 93–4, 96, 120–1
informed consent 74, 77–81, 121–2
  prospective 79–80
  retrospective 79
inside secondary analysis 12, 13
institutional review boards 76
intellectual resources 122–4
Internet 22, 91–2
interpretation 14, 62–3, 100, 123–4
interviews 27, 37, 41, 58, 95, 117

key terms
  electronic records 123
  literature search 127
knower-known relation 56

157

Labour Force Survey (LFS) 4
landscape of qualitative research
    16, 112–16
legal and ethical issues 25, 74–7, 81–2,
    103–4, 111–12
legitimation crisis 113
Lewis Terman longitudinal study 4
LFS *see* Labour Force Survey
life histories or stories 6–7, 21, 91,
    110, 129
literature
    review 126–30
    secondary analysis 1–2
longitudinal studies 3, 4–5, 9, 41, 46, 49

Mass Observation Archive 7, 21
material resources 119–21
member
    checks 66, 101–2
    passing as 101–2
    validation 100–2
meta-analysis 10–12, 15, 103, 122
meta-data 63, 65
meta-documentation 30–1, 63, 65
meta-ethnography 11
meta-research 11
methodological eclecticism 55,
    56–7, 111–12, 115
methodological matters 104, 122
methods criteria, literature evaluation
    129, 131–2
missing data 58, 94–5
modern forms of inquiry 113
modernist epistemology 123–4
modernist moment 112, 113
modes of secondary analysis 12–15
modi operandi 15, 89–108
moments of qualitative research
    112–13, 116
multiple data sets 48
    see also amplified analysis
multiple operationalism 100–1
Murray Research Centre data archive
    21, 83

narrative interviews 27
National Opinion Research Centre
    (NORC) 4
natural sciences 55
naturalist position 55, 56
naturalistic data 5–8, 15, 21, 37, 50,
    51, 55, 59, 68, 90, 96, 110, 117

naturalistic inquiry
    anti-foundationalism 100
    audit trails 68
    criteria 100
    positivism 66
    qualitative secondary analysis 100
    referential adequacy 69–70
    trustworthiness 100, 115
nature of qualitative research 54–7
nature of qualitative secondary
    analysis 110–12
negative case analysis 66
new or additional questions 8–9, 16,
    90, 110–11, 118
non-naturalistic data 5–7, 15, 37, 55,
    68, 96, 110, 117
non-naturalistic inquiry 5–6, 15–16,
    37, 110, 117
NORC *see* National Opinion Research
    Centre
not having been there 31, 60–5,
    102–3, 111–12, 115
notes 62, 117

omnibus qualitative data 7, 91
omnibus social surveys 3–4, 7, 30
on-line
    catalogues 22, 26, 91, 120
    data sets 91–2
    see also computing...; electronic...
open-ended questionnaires 50–1
operationalism, multiple 100–1
oral histories 21, 41
origins of data sets 122, 129
other researchers' data 14, 15, 31,
    61–3, 65, 90, 91, 95, 105
overview of studies 36–8
ownership
    copyright 84–5
    of data 73, 75

paradigm position 55, 57, 65,
    111–12, 115
participant observation 117
peer review guidelines 103–4, 131–4
personal data definition 85
personal secondary analysis 12, 13
phenomenology 55, 116
pooled case comparison 14, 48, 63–4, 98
positivistic position
    axioms 56
    conversation analysis 65–6

Subject Index

positivistic position *cont.*
  criteria 100
  ethnography 62
  naturalistic inquiry alternative 66
  qualitative research 55, 57, 62, 113
  quality assurance 100
  transcript transparency 67–8
  verification 65–6, 111
post-experimental inquiry moment 113
post-modern moment 113
post-modern position 113, 115, 123–4
pre-existing data, examples and
    types 2–8
presentation criteria 133
preservation of data sets 95
privacy rights 27
  see also confidentiality
professional associations 74–5, 78–9,
    81–2, 83–4, 90–1
protection
  confidentiality 82
  personal data 85
public records 3, 20

Qualicat 22, 26, 91
Qualidata
  confidentiality 82, 83
  copyright 84
  data archiving viii, 21–2, 26, 91
  data sharing 14, 114, 120
  Economic and Social Data Service 32
  ethical and legal guidelines 74, 76
  informed consent 79–80
  issues re-using archived data 10,
      61, 114
  UK Data Archive 21
qualitative data
  archiving 22
  computer-assisted analysis 99
  epistemological concerns 30–1
  methods of analysis 55
  naturalistic 5–6
  non-naturalistic 5–6
  secondary analysis 96–9
  sharing 28–31
  sources 110–11
  types of pre-existing 5–8
Qualitative Data Archival Resource
    Centre *see* Qualidata
qualitative research
  evaluation criteria 103–4, 131–4
  landscape 16, 112–19

qualitative research *cont.*
  perspectives 55–7
  traditions 54–7
quality
  assurance 99–102
  re-usability of data sets 93, 94–5
  research design 95
  semi-structured interview data 95
quantitative data
  access 26–7
  anonymization 82
  archiving 21–2
  modes of secondary analysis 13–14
  sharing 27–8
  sources 110–11
  types of pre-existing 3–5
quantitative secondary analysis 1–5,
    8–10, 12–13, 15–18, 110
questionnaires 40, 50–1, 117
quotations from transcripts 67

raw data 66–8, 70, 81
re-analysis 9–10, 30, 38, 45–7, 52,
    65–6, 111, 118, 129, 130
re-coding 98–9, 106
re-studies 46–7
re-using data
  assessing re-usability of data 92–6
  auto-data 63–5
  benefits and problems 26–31
  other researchers' data 61–3
realist position 55, 56, 57, 65
referential adequacy 66, 69–70, 111,
    115, 122
refinement function 9–10, 16
reflexivity 104–6, 123
refutation function 9–10, 16, 45
relevancy, data sets 91–2
reportage 102–4
reporting
  methodological matters 104, 122
  selective, primary research 42
  under-reporting 122
representation crisis 113
research diaries 63
research ethics boards 76
resources
  intellectual 122–4
  material 119–21
  technical 121–2
retention of data 22–6, 75
retrospective interpretation 14, 42

review of studies 35–53, 126–30
Roper Center for Public Opinion Research 20

samples adequacy 96
sampling 59
scratch notes 62
secondary analysis
    bricoleur 116–19
    cost-effectiveness 29
    definitions 1–18, 110
    functions 8–12, 16, 129
    future of 109–25
    landscape 112–16
    meta-research distinction 10–12
    modes 12–15
    nature of qualitative 110–12
    quantitative 1–5, 8–10, 12–13, 15–18, 110
    rationale 90
    see also data sharing; literature; types
selection of data sets 91–6
selective reporting 42
semi-structured interviews 58, 95
sequential analysis 58
Social Research Association 77–8
social surveys 3–4, 7, 19, 30
software 22, 99
sorting 59, 98
sources of data 12, 15, 110–11, 117
statistical data *see* quantitative data
suitability of data sets 93, 95–6
supplementary analysis 38, 41–5, 49–50, 52, 59, 79, 90, 111, 129, 130
supra analysis 38–41, 47, 48, 52, 59, 90, 98–9, 111, 118, 129, 130

symbolic interactionism 55
synthesis of research function 10–12, 16

tapes 95, 117
technical position 55, 56–7, 111–12, 115
technical resources 121–2
technician figure 116–17, 119
textual data 6, 82–3
    see also transcripts
theoretical sampling 97, 115
thick description 100
traditional moment 112, 113
traditions
    epistemological 114–15
    qualitative research 54–7
transcripts 22, 46, 66–7, 95
triangulation 46, 66, 100–1
trustworthiness 100, 115
truth claims 99–100
types
    pre-existing data 2–8
    qualitative data 55
    qualitative secondary analysis 35–53, 111, 129–30
    quantitative data 3
    quantitative secondary analysis 35–53

UK Data Archive 21–2, 76
under-reporting 122
universities 76, 84

validation 14, 45, 100, 101–2
veracity claims 100
verification 9–10, 16, 27, 30, 36, 65–71, 111, 115
vignettes 117